RFID for the Optimization of Business Processes

RFID for the Optimization of Business Processes

Wolf-Ruediger Hansen
AIM Global, Germany

Frank Gillert
UbiConsult, Germany

Translated by
Kenneth Cox
Kenneth Cox Technical Translations,
Wassenaar, the Netherlands

With a contribution from
Viola Schmid
Darmstadt University of Technology, Germany

John Wiley & Sons, Ltd

First published under the title RFID für die Optimierung von Geschäftsprozessen by Carl Hanser Verlag
© Carl Hanser Verlag GmbH & Co. KG, Munich/FRG, 2006

Authorized translation from the 4th edition in the original German language published by Carl Hanser
Verlag GmbH & Co. KG, Munich/FRG.

Copyright © 2008   John Wiley & Sons Ltd, The Atrium, Southern Gate, Chichester,
West Sussex PO19 8SQ, England

Telephone (+44) 1243 779777

Email (for orders and customer service enquiries): cs-books@wiley.co.uk
Visit our Home Page on www.wiley.com

Other Wiley Editorial Offices

John Wiley & Sons Inc., 111 River Street, Hoboken, NJ 07030, USA

Jossey-Bass, 989 Market Street, San Francisco, CA 94103-1741, USA

Wiley-VCH Verlag GmbH, Boschstr. 12, D-69469 Weinheim, Germany

John Wiley & Sons Australia Ltd, 42 McDougall Street, Milton, Queensland 4064, Australia

John Wiley & Sons (Asia) Pte Ltd, 2 Clementi Loop #02-01, Jin Xing Distripark, Singapore 129809

John Wiley & Sons Canada Ltd, 6045 Freemont Blvd, Mississauga, ONT, L5R 4J3

Wiley also publishes its books in a variety of electronic formats. Some content that appears in print may
not be available in electronic books.

British Library Cataloguing in Publication Data

A catalogue record for this book is available from the British Library

ISBN 978-0-470-72422-4

Typeset in 10/12pt Optima by Laserwords Private Limited, Chennai, India
Printed and bound in Great Britain by TJ International, Padstow, Cornwall
This book is printed on acid-free paper responsibly manufactured from sustainable forestry in which at
least two trees are planted for each one used for paper production.

To Ute and Marita

Contents

Contributor

Professor Viola Schmid PhD, LLM (Harvard) assumed the Chair of
Public (International) Law in the Faculty of Law and Economics of the
Darmstadt University of Technology (Darmstadt, Germany) in 2002.
Her research areas are cyberlaw, e-justice and freedom of speech.
Contact: schmid@jus.tu-darmstadt.de.

Foreword

Radio frequency identification (RFID) is used in countless areas where there is a need for wireless identification of objects and transmission of data. RFID devices can take the form of wireless tags attached to products, cases or pallets, or they can be implemented in smart cards or telephones with smart card attributes carried by ordinary citizens. It is foreseeable that RFID technology will spread over the entire world. Companies in the USA and Asia are committed to rapid introduction of this technology, while Europe is still hesitating.

Naturally, we must not ignore the privacy risks associated with using RFID technology, to the extent that data carried by or associated with wireless tags can be used to deduce the behaviour of citizens or obtain information about them. There are still many unresolved issues here, which also relate to legal aspects. European legislators have not yet arrived at a definitive position on the use of RFID technology. For this reason, the European Parliament launched an initiative on the theme of RFID in early 2006, under the umbrella of the Scientific and Technological Options Assessment (STOA) programme.

Viviane Reding, European Commissioner for Information Society and Media, opened the debate on 9 March 2006 at the CeBIT fair in Hanover. With this, she highlighted the importance of RFID, including its significance for the Metro Future Store Initiative. Commissioner Reding was feeling her way toward a concrete, written legislative initiative by conducting expert hearings. In May and June 2006, Key Theme workshops for this purpose were held in Brussels under the direction of the EU, with numerous experts among the participants. In March 2007 at the CeBIT again, Reding presented a Communication addressing major RFID issues. In June 2007, the Commission organized a conference in Berlin titled 'Toward the Internet of Things', which was widely regarded as very successful.

The aim of the STOA activities initiated by the European Parliament is to discuss a visionary approach and develop future scenarios that reflect the perspectives of industry, suppliers of technologies and services, and consumers. This is intended to give the members of the European Parliament and decision-making bodies at the European level a clear understanding of how they can help foster European interests in RFID technology and secure market opportunities for Europe while at the same time protecting citizens against abuse of this technology. In early 2007, STOA presented a complete study on the anticipated consequences of RFID technology. This study is available on the STOA website.

In parallel with this, measures for Research Programme 7 (RP7) must be developed to provide successful incentives for research and industrial development on the way to creating the 'internet of things'.

I am pleased to see that this book addresses not only the technical aspects of RFID technology, but also the application aspects – in other words, the technology as seen from the perspective of industrial, logistical and consumer-related processes that can be enhanced with RFID. This helps keep the debate objective, and the many application scenarios show that RFID technology is already used much more extensively than most laypersons realize. This book thus comes at the right time to underpin the European discussion and clarify the technology for future users.

I hope that a large number of people read this book, and I expect that it will give the debate an objective point of reference.

Jorgo Chatzimarkakis,
Member of the European Parliament
Brussels

Preface

This book is aimed at two groups: organization specialists and people who are responsible for optimizing business processes. We wish to assist them in developing concepts for improving business processes that are intended to be implemented using IT systems and radio frequency identification (RFID) systems. We also describe the main features of the underlying technical infrastructure of RFID technology in order to give organizers an overview. Nevertheless, we recommend that technology experts consult more detailed literature, such as the proven *RFID Handbook* by Klaus Finkenzeller, which is also published by John Wiley & Sons.

Our primary objective is to forge a link between the technology and the application scenarios for using the technology to support business processes in business and industrial environments. Here we believe that it is especially important to show that the benefits of RFID methods are not limited to the latest generation of RFID tags, but instead result largely from their potential uses in processes, in particular logistics processes. The task facing potential users is thus to determine the potential benefits of using RFID technology in their enterprises, or better yet do this in cooperation with their business partners, in order to compare them with the associated costs of the envisioned solution (and in this order). RFID technology applications often fail due to lack of agreement on how to allocate the costs and benefits of processes that span enterprise boundaries. We devote considerable attention to the business processes concerned in order to help overcome this hurdle.

The public debate on RFID technology is presently focused too much on the details. For instance, some people claim that the only thing necessary for the commercial success of RFID tags is a sufficiently low price. Others hope that all that is necessary is to refine the antennas of RFID tags sufficiently to obtain the required reading properties. These discussions focus too much on individual technical aspects and fail to see the overall picture. Improving the edgeware, which consists of the software components that control the readers, is also only one part of the overall picture.

By now, tags with unit prices of less than 10 eurocents have been announced, but there is still no explosion in use of RFID technology. Why not? Because the next step on the way to the global use scenario still has to be taken: installing a broad base of RFID readers and integrating them into enterprise resource planning (ERP) systems, which run on corporate servers and perform core business management tasks. In parallel with this, communication systems must be extended along the logistics chains of the enterprises concerned. Tag prices are only a small part of the overall picture.

Current market trends also show that there will not be any single, common RFID technology for all application scenarios, because the physical laws that govern the behaviour of radio signals create a variety of obstacles. These obstacles can only be overcome by using a variety of RFID technologies. Here we particularly have in mind the capabilities and limitations of the various frequency bands. RFID systems can operate in selected frequency bands between 125 kHz and 5 GHz. Although the efforts of industry consortium EPCglobal to establish the UHF band (860–960 MHz) as the only global standard in the trade sector may have succeeded with regard to RFID tags for cases and pallets in trade environments, the frequency of 13.56 MHz (HF band) will be more significant for wireless tags on individual articles. Reader manufacturers are already developing reader antennas that can handle both frequencies. EPCglobal has recognized this situation and announced that the Generation 2 standard, which up to now has been based on the UHF band, will be extended to include the HF band.

In the European debate on RFID, we often have the impression that market penetration has already progressed significantly further in the USA than in Europe. In our view, this impression is false. Among other things, it relates to the fact that the level of organization and extent of IT penetration in logistics chains are lower in the USA than in Europe, which means that the US situation offers significantly more room for improvement by using RFID technology and other measures. As a result, many RFID success stories in the USA can only partly be attributed to the use of wireless tags, with general process improvements and increased employee discipline also playing a significant role. This has been shown by a recent study carried out at Wal-Mart [Hardgrave05].

The claim that barcodes are on the verge of being replaced by RFID, which is made repeatedly in euphoric marketing statements, is actually counterproductive and false. Both technologies will persist in a complementary relationship for many decades. It is thus the duty of planners in organizations to comprehensively assess the benefits of all auto-identification technologies and assign each technology the tasks for which it is best suited. Helping them to do this is also one of the aims of this book.

This book would not have been possible without the information and assistance the authors received in conversations with numerous experts. Our discussions with Professor Elgar Fleisch and Professor Friedmann Mattern were especially valuable. Professor Fleisch is a director of the Institute of Technology Management at the University of St Gallen (HSG) in Switzerland and co-chair of the Auto-ID Labs (www.autoidlabs.org), where he is working with a global network to develop and refine the infrastructure of the 'internet of things'. Professor Mattern lectures at ETH Zürich and is head of the Distributed Systems Research Group of the Department of Computer Science. The two professors jointly head M-Lab (www.m-lab.ch).

The seed for this book was sown by the RFID White Paper published in Berlin in 2005 by the RFID Project Group of the industry association BITKOM eV. The authors of this paper collaborated with experts in major IT enterprises, including Cisco, GS1, Hewlett-Packard, IBM, Infineon, Intel, SAP, Siemens and T-Systems. We are grateful to BITKOM for permission to utilize their insights in this book. We also extend special thanks to Dr Norbert Ephan of Kathrein-Werke KG in Rosenheim, Germany, who never tired of adapting our process-oriented requirements to the engineering realities and physical circumstances that govern the use of RFID tags, readers and antennas.

We also owe a special word of thanks to the EICAR RFID Task Force, whose theme 'RFID and data protection' provides the basis for Chapter 8. The chair of the task force, Robert Niedermeier, was kind enough to proofread the original text and confirm his reputation for legal correctness.

The German legislative perspective described in chapter 8 is enlarged in chapter 9, which presents a global overview of RFID legislation in various countries throughout the world. We owe special thanks to Professor Viola Schmid, Chair of Public (International) Law in the Faculty of Law and Economics of the Technical University of Darmstadt, Germany, who provided chapter 9 specifically for the English edition of our book. This substantially increases the value of the book for an international readership.

Furthermore, many thanks go to Peter Kreuzer, an automotive expert of the VDA, who provided us with a deep insight into standardization in the automotive industry and thus enabled us to include a section on this market sector.

Finally, we are proud of two contributions from the academic world: the description of agent technology in internal logistics systems (Section 6.2 and the application case in Section 10.12) by Dirk Liekenbrock of the Fraunhofer Institute for Material Flow and Logistics in Dortmund, and the description of the requirements for implementation of RFID infrastructures (Section 7.4) by Jan Hustadt of the Logistics Department of the University of Dortmund.

We would also like to thank the companies that enabled us to document the application cases described in Chapter 10. They show that proliferation of RFID use occurs unobtrusively more often than openly, and that it is worthwhile for process managers in enterprises to profit from the experience of others by looking for existing solutions that can serve as models and be adapted to their own needs.

Last but not least, we would like to express our appreciation for the patient, trustworthy and constructive cooperation of Margarete Metzger and Irene Weilhart at Carl Hanser Verlag and Simone Taylor at John Wiley & Sons, and the translation by Kenneth Cox. They gave this book its final and vital finishing touches.

Dr Frank Gillert and Wolf-Ruediger Hansen
Frankfurt/Munich

1

Introduction

Radio frequency identification (RFID) is a seemingly simple technique. Data is stored in RFID tags that are attached to objects or located in smart cards, and this data can be read using radio signals and presented on a display by using a suitable reader. The data can then be transmitted automatically to an information technology (IT) system for further processing. Although this method is easy to describe in technical terms, there are many obstacles in the areas of application technology and integration into operational processes that must be overcome before it can be used operationally. For this reason, this book concentrates primarily on the potential economic benefits that can be realized from improvements to business processes that can be achieved using RFID, rather than on the technical aspects of how RFID tags and readers work and what they do. Our objective here is to examine RFID technology as a whole in the context of enterprise processes and higher-level IT systems.

There are essentially two groups that are interested in using RFID technology. The first group consists of the innovators, the advocates of new technology, who argue that it can create added value in the form of short-term or long-term benefits. They incorporate RFID technology in their strategic plans and start experimenting with it at an early stage. Other enterprises, by contrast, find themselves forced to introduce RFID because their customers demand it, for example due to the policy of mandated requirements pursued by large retail groups. In addition, they are afraid of losing ground to competitors and missing out on competitive advantages. In terms of marketing theory, the first group is referred to as the innovators, visionaries or early adopters, while the second group is referred to as the 'conservative majority' or 'laggards'. They are also called the 'fast' and 'slow' groups [Moore96].

In this book, RFID is regarded as one of several auto-ID technologies that can be used to identify objects or persons. It is not our intention here

RFID for the Optimization of Business Processes   Wolf-Ruediger Hansen and Frank Gillert
© 2008 John Wiley & Sons, Ltd

to create the impression that RFID is always the ultimate solution. We are convinced that barcodes still have a long life ahead of them and that they will be used alongside RFID in a complementary fashion, particularly in trade environments.

The debate on the potential success of RFID technology is not aided by the fact that a global media hype has developed on the subject of RFID. For one thing, this hype fosters fear of losing control over how personal data is used and reduces willingness to accept new technologies. It also encourages enthusiasts to devise unrealistic scenarios and lose sight of circumstances that constrain the use of the technology, in particular physical constraints. However, euphoric behaviour of this sort is often seen when a new technology is introduced. A US marketing research company, Gartner, has developed a model of the hype cycle in order to describe this phenomenon. The RFID hype cycle is described in Chapter 2, 'Visions, Reality and Market Drivers', which also shows that there are many ways to put RFID technology to good use, even if some approaches ultimately land in the 'trough of disillusionment' as often happens with innovative technologies.

If you follow the public statements of the industry consortium EPC-global and its members – e.g. retailers such as Wal-Mart and Metro – you often get the impression that RFID is already in common use. However, this is not so. Here it is important to recognize the correct context for viewing the various aspects of RFID applications. Up to now, the key significance of the RFID innovation project of the Metro Group lies less in the actual implementation of RFID methods than in their comprehensive portrayal of the capabilities that could be made available to retailers in the medium to long term, not only in branch outlets but also along the entire supply chain. With its commitment to RFID, the Metro Group has created an internationally communicated image as an innovator. In the view of the international market, this puts it at the same level as much larger groups such as Wal-Mart in the USA. This should be regarded as an aspect of corporate strategy, and in particular as a way to send a message to the stock markets. Information about the status of the project and the accumulated experience and insights gained from it are reported quite extensively in Metro's freely available *RFID Newsletter*, which appears several times a year [Metro2603].

Another factor that blurs the issue, and which is an unintended result of the extensive publicity efforts of the EPCglobal consortium, is the fact that EPCglobal's market activities are often regarded as generally representative of all possible RFID solutions. They should instead be seen in perspective, because there are many areas – such as production control systems in manufacturing companies and smart card applications – that lie outside the sphere of influence of EPCglobal. Anyone interested in the subject is well advised to acquire a good understanding of the various aspects and players in order to obtain a balanced view of the

RFID situation. The applications described in Chapter 10 also provide a suitably broad overview.

In the EPCglobal domain, which means in the retail sector and other directly related sectors, practical implementation of RFID techniques is proceeding much more slowly and with much more effort than what can be seen at first glance. In the press and at trade fairs, attention is often focused on individually tagged items. We see them being placed in special shopping carts, which must be made from plastic so the antennas will not be screened by the metal grids normally used for this purpose. Then we see how the items are automatically identified at the checkout and a receipt is generated 'on the fly'. This scenario is still a fantasy, as according to Metro's own statements it will take at least five years before comprehensive item tagging becomes generally established. As an exception to the situation in the EPC world, clothing articles are increasingly being tagged at the item level. In the retail sector, the primary interest at present is on tagging pallets and cases, which addresses logistics issues. Outside the retail sector, one example of an application where RFID tags are already being used successfully is in identifying books. The current process drivers for RFID technology are described in Chapter 2.

Chapters 2, 3 and 4 are largely devoted to logistics aspects, since this is where the greatest potential benefits are expected to be found. However, the situation is complicated by the fact that the potential benefits can only be realized if the various enterprises in a particular supply chain can agree on comprehensive solutions. Consequently, decision-makers in these enterprises need more support with projects that involve using RFID in logistics chains – which means planning 'open' systems – than with projects that involve using RFID to support internal processes ('closed' systems), such as in production control environments.

One of our specific objectives in describing logistics aspects is to help small and medium-sized enterprises (SMEs) understand the process-specific context so they can start participating in the RFID processes of the major players in a timely fashion. For their part, the large retail groups must include SMEs in the process if they want to see their RFID strategies realized in the end. They will pursue this goal by using more or less gentle persuasion ('mandated requirements'). For players in the SME sector, it is thus important to be prepared for the introduction of RFID in order to avoid the risk of losing customers due to insufficient technological adaptability.

The potential benefits that can be seen in the market are discussed in detail in Chapter 3. There we describe the market structures, value chains and foreseeable market consolidations in order to give users a better understanding of why suppliers act in particular ways and the strategies they develop to position themselves in the market.

Chapter 4 presents a highly detailed view of planning and process structures, which above all will force supply relationships to change from the existing pattern of bilateral relationships between suppliers and customers to a pattern of multilateral network relationships – the 'supply net'. Using RFID technology is also expected to eliminate a well-known shortcoming of the retail supply chain: the 'bullwhip effect'. This effect occurs because the volumes of merchandise sold in branch outlets and held in intermediate stocks or by suppliers are systematically overestimated or underestimated. In the first case, this generates excessive stock levels, which tie up significant capital and always lead to asset losses, for example when goods can no longer be sold because they have become outdated in stock. In the second case, it results in stock shortages that lead to empty shelves (out-of-stock items). The net result is a combination of avoidable costs and lost revenue, which can easily consume an already narrow profit margin.

We also discuss planning and process structures that can be supported especially effectively by RFID methods and can be utilized either incompletely or not at all in the absence of RFID, such as vendor-managed inventory (VMI), efficient consumer response (ECR), and collaborative planning, forecasting and replenishment (CPFR). As can be seen from this discussion, enterprises that cannot agree with their trading partners on transparent management of supply chains are also not sufficiently mature to use RFID technology. Based on these general approaches, we then examine the potential benefits in more detail. We discuss suppliers in the packaging industry and service provider structures. This is followed by an in-depth look, at the subprocess level, of analytical models for assessing economic viability.

Data processing in IT systems always involves identifying the objects for which the data is to be processed. Introducing RFID leads to serialization of the objects, which means that individual objects can be identified uniquely. The barcodes presently used in merchandise systems only allow the product type to be identified. A barcode consists of a manufacturer number (producer) and a product number (object type). Schemes such as the Electronic Product Code (EPC) can be introduced with RFID systems. This code includes a serial number in addition to the other data. This makes it possible to distinguish one bottle of apple juice from another one next to it – every bottle has a virtual identity. The EPC strategy is the focus of Chapter 5.

In the case of apple juice, it might reasonably be asked whether this serialization is worth the effort. However, it becomes a lot more sensible if you consider medicine packaging. Besides the logistics benefits, one of the objectives here is to protect products against counterfeiting, and particularly in the pharmaceutical sector to maintain pedigrees (e-pedigrees) that describe a product's history from production through the entire retail supply chain to the consumer. Here pharmaceutical products must be handled in the same way as technical replacement parts for cars or aircraft. Pedigrees also allow recall campaigns to be conducted more specifically and efficiently than is presently possible. This yields benefits for consumers as well as enterprises. For all of this to be possible, it is necessary to develop standards that accelerate global participation of enterprises in using RFID technology. These aspects are also discussed in Chapter 5.

The 'internet of things' is a complex structure. In order to understand its manifestations in public IT structures and IT structures inside enterprises, it is helpful to understand how to interpret architectural terms in the IT context. Chapter 6 is dedicated to this objective. Based on the historical development of IT architectures, we show that many structural elements necessary for novel RFID applications are actually not all that new. Seen from this perspective, RFID is simply a medium for transmitting relevant event data necessary for handling operational processes directly and automatically in IT systems. After the data has been input into the system, further processing of the data does not depend on the technology used to acquire the data. However, data processing and database updating can be performed closer to real time with RFID. Mapping of real-world business processes into IT systems is thus more realistic, and IT systems can do a better job of performing their control tasks.

Agent technology provides IT architectures with a completely new, decentralized structural element. Agents are independent software modules that can be used in local or mobile environments. They are autonomous instances that become active without centralized control by IT systems. Agents can be used in the edgeware domain, in readers or even in the actual objects if they are equipped with suitable processors. As an example that illustrates the role of agents and how they work, we describe an industrial conveyor system that independently determines successive destinations for routing individual containers along a branched conveyor belt system according to the necessary processing steps or available capacities. The agents in this system also communicate with each other via RFID.

A paradigm shift with regard to operational information systems is currently taking place in this area, and it requires new ways of thinking on the part of system managers. Up to now, operational information systems have been implemented using centralized approaches. Enterprise resource planning (ERP) systems, and in particular the systems supplied by market leader SAP, are perfect examples of this. By contrast, agents perform control tasks independently and thereby decentralize process control. RFID technology also has a decentralizing effect even without introducing agents. This can be seen from the increasing availability of supply chain management (SCM) software packages that run in the edgeware domain close to RFID readers instead of in the background on a mainframe computer as in the past. It is thus advisable for anyone involved in innovative use of RFID technology to acquire an understanding of these aspects of decentralization.

Chapter 6 concludes with a description of on-demand services. This is because many enterprises, especially those in the SME sector, are more likely to yield to the pressure of innovation than to entrust operation of their IT systems, either entirely or in part, to external service providers. This is usually called 'outsourcing', but the term 'on demand' is becoming increasingly common in connection with this form of service provision. It is intended to convey the idea that a good service provider always supplies exactly the amount of IT support needed by the customer organization, and, more importantly, only charges for this amount of support.

In Chapter 7 we turn our attention to the hardware infrastructures of auto-ID and RFID systems. We are convinced that it is helpful to understand the basic features of the technical structures of RFID tags and reader antennas and the processes used for communication between them. An especially critical aspect here is the 'air interface', which is the electromagnetic field between the antennas of the tags and readers that is used to transmit data. This field is governed by unyielding physical laws that must be respected by enterprises when they are planning RFID systems. Reader manufacturers often give the impression that it is only a matter of time before their readers reach a level of technical sophistication that enables them to resolve every reading problem. For their part, producers of RFID inlays (combined chip/antenna modules used in RFID tags) suggest that they will soon discover the mythical philosopher's stone by fashioning the antennas (in this case dipole antennas on UHF inlays) in the form of especially imaginative fractal shapes. Although improvements can be achieved in all areas here, it must be borne in mind that the read fields of the antennas are subject to physical laws, with the result that disturbances by metallic objects or even liquids can make operational use impossible due to inadequate read reliability. Simply rotating a case by 90° can cause its transponder to fall outside the range of the antenna – not because the distance is greater, but because the resulting 90° angle between the RFID tag antenna and the reader antenna reduces the effective range. Relevant results obtained from experimental systems are described in Chapter 7.

The evident technical limitations on process optimization must be compensated by using process-specific measures. In particular, requirements for attaching tags to packages and arranging packages in containers or on pallets must be specified precisely, and these requirements must be obeyed. In this regard, RFID methods can also fail due to a lack of employee discipline or inadequate training. Section 7.4 deals with these requirements for successful RFID implementation.

In Chapter 8 we highlight the essential statutory provisions with regard to consumer protection and data protection in Germany. However, we are convinced that the vast majority of operational application scenarios for RFID have little or no relevance to consumer protection. In the retail merchandising environment, special caution is of course necessary if comprehensive labelling with RFID tags is ever to become generally established. However, this is still several years away, as already mentioned. Consumer-oriented near-field communication (NFC) applications, which are discussed in Chapters 7 and 10, could spread more quickly. Consumer protection is naturally very important in this area.

Some fear scenarios with regard to inadequate data protection in RFID applications can easily be put into perspective. They arise in part from inadequate communication by enterprises regarding the use of RFID tags, and in part from ascribing capabilities to the technology that it simply does not have. For instance, it is quite difficult to even read RFID tags attached to articles of clothing without this being noticed, and even if this can be done, it is difficult to do much with the information because the tags usually contain only item numbers. Data stored electronically in e-passports can only be recognized after the text information has been read optically.

Attempts to store large amounts of data in tags should be met with scepticism. All information about objects or persons that can be identified using RFID tags or smart cards can be made available in databases accessible via the Internet. This data will be collected there anyhow, independent of the use of RFID. Storing a large amount of data in a RFID tag prolongs the read process, which is a disadvantage in the logistics environment.

Chapter 9 extends Chapter 8 by presenting a global overview of RFID legislation in the USA and other countries based on recent research activities at the Technical University of Darmstadt, Germany. It was written especially for this English edition of the book by Professor Viola Schmid, whose research areas are cyberlaw, e-justice and freedom of speech. An interesting point here is that the US Food and Drug Administration has extended its view beyond RFID to include nanotechnology, encryption technologies and other methods.

Finally, in Chapter 10 we invite the reader to join us on a tour of applications that have already been implemented or will be implemented in the near future. In that chapter, as in all others, our aim is always to encourage readers to let the contents stimulate their imaginations and create links to similar circumstances or situations in their realm of experience and professional environment.

In addition to the list of references, the appendix includes a glossary of the most important terms and abbreviations and a directory of Web addresses intended to help the reader learn more about RFID.

2

Vision, Reality and Market Drivers

If you don't have patience with the details, you will cause your major projects to fail.

– Confucius

In this chapter, we take stock of the market situation and explain the mechanisms that have led to the major media hype around RFID. We also explain which visions appear to be realistic, which generic potential uses can be expected and which market drivers are at work.

The primary aim of using RFID technology is to reduce the transaction costs of operational processes at the interface between the real world and the virtual world. These two worlds are presently separated by a relatively large gap, which is also called a 'media break'. To close this gap, real business processes and their virtual representations in information technology (IT) systems must be reliably aligned and linked to each other. The media break is depicted in Figure 2.1 as a horizontal funnel along the time axis. It lies between the information systems located the upper region and the real processes located in the lower region.

This funnel narrowing to the right represents:

- reducing the cost of data acquisition;

- reducing the time between data generation and availability in IT systems;

- increasing data quality.

The 'classic' data acquisition route via keyboard entry from paper forms generates the highest costs and produces the longest delay in updating databases. Problems due to acquisition errors, including human

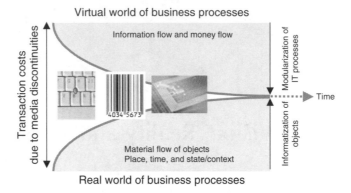

Virtual world of business processes

Real world of business processes

Figure 2.1   The media break between the virtual world and the real world (based on [Flei2005])

errors, and problems resulting from initiating incorrect control processes because the information in the information system databases is usually not up to date should also be classed among the costs. Data can be acquired faster and more reliably by using barcodes. However, manual support is still necessary, for instance to give the reader a line-of-sight view of the barcode. RFID methods reduce costs even further because RFID tags can be read through packaging materials.

Real business processes are primarily related to physical material flows of objects. Location and time data can be acquired fully automatically using RFID methods. Arrival of such information at an RFID reader is called an 'event' or an 'identification event'. Status information (e.g. an item is damaged) or context information (e.g. an item is part of a load unit, such as a pallet) regarding objects must be entered manually if necessary. With RFID, full automation of event processing is possible in the medium term. This will shift the situation to the tip of the funnel and eliminate the media break.

In specialist jargon, the terms auto-ID (automated identification), AIDC (automated identification and data capturing) and ADC (automated data capturing) are used to describe automation of data generation. These terms encompass a variety of methods, including commercially established basic technologies for data acquisition at various levels of automation. These technologies, methods and products are being developed by numerous enterprises throughout the world. Many of the market players in this sector are members of the Association for Automatic Identification and Mobility (AIM),[1] an international consortium represented by local organizations in most industrialized countries. The colloquial term auto-ID is becoming increasingly common in general use as a short form for 'automatic identification'.

---

[1] For information about AIM, see www.aimglobal.org (international) or www.aimuk.org (UK).

There are many different auto-ID methods, all of which are based on four basic physical principles [Arn95]:

- **Mechanical and electromechanical identification systems** use mechanical components such as cams, probes, sheet-metal tabs or contacts as binary information storage elements or use capacitive or inductive sensors for reading.

- **Magnetic identification systems** use the magnetic fields of permanent magnets or magnetic coatings on magnetic cards, magnetic stripes or magnetic ribbons (comparable to magnetic stripes on credit cards) to store information. Reading requires close proximity and very precise guidance.

- **Optoelectronic identification systems** recognize object outlines or applied markings such as colours, reflective marks, OCR fonts or barcodes.

- **Electronic and electromagnetic identification systems** are based on electronic data storage media. Data is transmitted by electromagnetic waves or induction without physical contact. Pre-programmed and freely programmable storage media are available, usually based on microchips. The storage medium can be designed to be active (with a built-in battery) or passive. In the first case, the built-in energy source is used for communication, while in the second case energy taken from the electromagnetic field of the reader is used for this purpose. The following terms are used in common practice: mobile data storage device, programmable data storage device, RFID tag, transponder, and simply RFID (radio frequency identification).

Conventional auto-ID technologies are still in use, but the trend is toward optoelectronic and electromagnetic systems, although barcodes in particular will continue to be used in parallel with RFID for many years because they are less expensive and functionally adequate in many cases. Not surprisingly, the auto-ID method used in any given case is selected based on cost/benefit considerations [Flei2005].

In brief, it can be noted that RFID is one of many possible auto-ID technologies. However, it is the only technology that has the potential to develop into a separate class of decentralized IT infrastructure components, which is also described as 'informatization'. It would be possible to 'informatize' objects in a material flow by storing information such as packing slips or product instructions in their RFID tags, which could be read out at any desired time by a centralized goods management system without physical contact. The next step would be to equip the RFID tags with their own processors so they could independently perform logical processing. Among other things, this forms part of what is called 'agent technology'. This trend is also fostered by the availability of sensors with constantly decreasing prices and dimensions and constantly increasing capabilities.

All this makes fully automated data generation inside the process chain possible under increasingly economical conditions. This will lead to a continuous process of information exchange, which with suitable transmission protocols and database structures will result in objects having an 'information presence' comparable to the Internet. This is also an aspect of the 'internet of things' [Flei2005].

The transformation from discrete data acquisition to continuous, event-based data acquisition is accompanied by a trend toward increased modularization and networking of business process support software and running this software on platforms close to the objects. RFID thus helps narrow the gap between the real and virtual worlds, as indicated by the funnel in Figure 2.1. We can expect to see a steady decline in data transmission costs and continuous improvement in the consistency and currency of virtual systems relative to the real world.

Naturally, the significance of RFID is also reflected by its media presence. For instance, Figure 2.2 shows a sharp rise in the number of mentions of RFID in the area of retail logistics relative to media coverage of conventional supply chain management techniques. This high mention rate often gives the impression that RFID overshadows all other techniques, which is usually a distortion of actual management

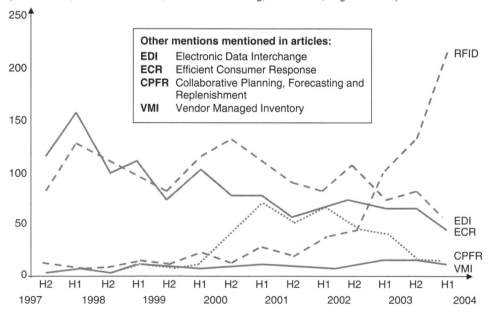

Figure 2.2   Media presence of current supply chain management concepts (source: *Logistik Heute*)

potentials. This is the negative side of the hype. The other techniques[2] shown in Figure 2.2 are still significant in their own right, are partly supported by RFID, or have nothing at all in common with RFID. A figure such as this thus gives the wrong impression if it is not accompanied by conscientious commentary.

The term 'efficient consumer response' (ECR) is used in the retail world to designate an overall strategy for properly meeting consumer demand. ECR comprises two organizational components: efficient replenishment (efficient shelf stocking) and category management (appropriate arrangement and presentation of goods). Methods for cooperation between suppliers, logistics service providers and retailers have been developed to ensure that these processes run smoothly. The best-known methods are collaborative planning, forecasting and replenishment (CPFR) and vendor-managed inventory (VMI). Near-real-time exchange of information between the enterprises linked together by the supply chain is essential for implementation of the above-mentioned concepts. Various types of processes for this purpose have become established in various market sectors. They are collectively referred to as electronic data interchange (EDI). Orders, packing slips, receiving confirmations, invoices and other electronic documents are transferred between trading partners via EDI.

An operational EDI process is a prerequisite for effective use of RFID. This can be seen by examining what happens at the loading dock of a trading company's distribution centre. This is where the pallets with RFID tags arrive. The packing slip number or serial shipping container code (SSCC) is read by an RFID reader. The IT system then retrieves the packing slip, which was previously transmitted by EDI, so the received goods can be checked and forwarded. If the packing slip is not available, the RFID tag does not provide access to more comprehensive information. This may change in the future when tags have enough storage capacity to hold an entire packing slip. However, RFID technology has not advanced this far yet, especially considering that the reading process must be completed very quickly. The more data that is stored in the tag, the longer the reading process takes.

The term 'RFID' has come to have a somewhat independent existence these days. The impression is often created that RFID is an independent megaconcept that will render all other methods obsolete. The above example of collaboration between RFID and EDI shows that this is not the case, or at least not within the foreseeable future.

RFID technology is often judged, especially by contemporaries enamoured of technology, in terms of a visionary remote goal of a conceivable migration path. This creates expectations for rapid, far-reaching realization of the potential of RFID, which in fact can only be achieved step by

---

[2] The methods mentioned here are described in detail in Chapter 4.

step over a very long period. Unrealistic expectations can quickly lead to frustration. This is the essential reason for the negative aspects of the hype around RFID. These expectations must be tempered by reality-based, economically sound information. People must be properly informed about the practical steps that will lead to the future of RFID technology.

The 'hype cycle for emerging technologies' devised by US market research company Gartner[3] portrays the progress of new technologies from commercial launch to maturity, plotted against a time scale, for technologies that have emerged since 1995 (see Figure 2.3). The vertical axis shows qualitative phenomena, in particular the initial surge in the expectations of market players and an increasing level of market penetration and economic substance or degree of maturity of the technology.

The hype cycle always has the same typical shape. Specific technologies and their estimated expectation levels are plotted on this curve. It starts with a sharp rise in expectations until the 'peak of expectations' is reached. Up to this point, the technology has achieved a certain level of market penetration because innovative enterprises that are willing to take risks have seen opportunities in it. Then the first failures start to occur. As is generally known, negative publicity is much more effective than positive publicity. As a result, the technology comes into disrepute relatively quickly. The initially positive expectations flip over into frustration. Many market players cancel their projects and turn elsewhere. This causes the curve to drop into the 'trough of disillusionment'.

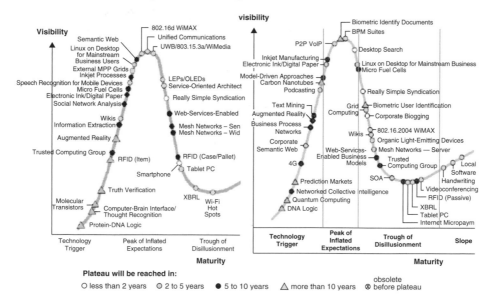

Figure 2.3   Comparison of the RFID hype cycle in 2004 and 2005 (source: Gartner Group)

---

[3] Gartner is an international leader in research and analysis of the international information technology industry (www.gartner.com).

Nevertheless, the determined and realistic players among the users discover the potentials that lead to the desired goal and use the technology in a manner that generates benefits for them. These cases are shown on the portion of the curve rising out of the trough of disillusionment, which is called the 'slope to maturity'. However, successes along this path are often not communicated by enterprises. Companies are afraid of giving their competitors insight into how they have improved their cost structure or competitive position by suitable use of technology, and they want to enjoy the benefits of their advantage.

Even if reports on successful applications appear in the media, this news does not receive nearly as much public attention as the hype messages, whether negative or positive, during the first part of the hype cycle. This can also be seen from the fact that the peak of expectations is much higher than any level that will ever be reached on the slope to maturity. The phenomenon is thus called 'hype' due to the exaggerated media attention.

Figure 2.3 shows the position of RFID on the hype curves in 2004 and 2005 for comparison. Item tagging and case/pallet tagging are shown separately on the 2004 curve. Item tagging was in the middle of the upward slope of the expectation curve at that time. It is no longer shown separately on the 2005 curve. We now know that it will take at least five years before it becomes available in the retail sector. By contrast, it is already gaining acceptance for technical replacement parts. In order for the hype curve to present this level of detail, it would necessary to assign individual points to the various application areas. For 2005, the only category listed is 'RFID (passive)'. However, it is already starting to move up the slope to maturity, which is an indication that the major retailers are receiving more and more pallets with passive RFID tags.

The Gartner hype cycle is strictly intended to be regarded as a methodical portrayal of a generally observed phenomenon. More advanced, substantiated analyses of technology development are carried out using the method of technology management, which differentiates between categories of technology maturity in order to structure the technology discussion. The following analysis is intended to show the extent to which comprehensive classification of all conceivable technology variants and applications under the name 'RFID' leads to further misinterpretations.

The following definitions of technology categories form the basis for further analysis [Speck2004]. They clearly show a close methodical relationship to the Gartner hype cycle.

- **Future technologies** are in the research stage in universities or the basic research centres of enterprises.

- **Pacemaker technologies** are still in the early development stage, but the first specific application areas have already been identified.

- **Key technologies** are on the verge of becoming basic technologies, but they are still in an intensive development stage. They harbour strong potential for innovative developments in the process and product domains.

- **Basic technologies** is the name given to generally used technologies that are already in the mature phase. No entrepreneurial risks are associated with using these technologies.

Figure 2.4 shows these technologies plotted against a time scale. The more experimental a technology still is, the greater the potential for change ascribed to it from the perspective of technology management.

It is astonishing that RFID has already become a basic technology in so many application areas without being perceived as such in the public awareness. For example, remote control of central locking systems in cars is an RFID technique that every driver uses daily. Another example comes from vehicle manufacturing, where the product carriers that transport unfinished chassis between assembly stations are fitted with RFID transponders containing complete production order data in the transponder memory, which can be read out at every assembly station.

By contrast, the public debate often creates the impression that RFID applications such as attaching tags to retail items are still standing in the wings and will not materialize until many years from now.

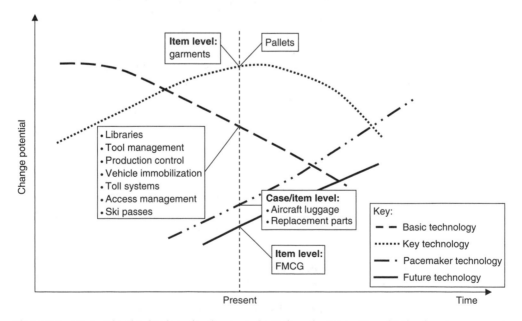

Figure 2.4   Time scale of technology development phases from the perspective of technology management

A scenario that is often portrayed by retail organizations, with a shopping cart filled with items that can be identified automatically by a RFID reader at a checkout with the sales slip being generated as the cart is pushed past, is a figment of the imagination for the foreseeable future and counterproductive to enlightenment about RFID. Systems have not advanced this far yet because tags are not yet cheap enough and the reading speed of antennas is not yet sufficient to read the contents of shopping carts or even pallets full of tags.

The art of discovering successful applications lies in distinguishing what is possible from what is not and differentiating between short-term and long-term aspects. Here negative experiences are often plucked out and portrayed in exaggerated form to justify blanket rejection of a new technology or escape the need to make a decision about using the technology. This is also referred to as a 'don't-care situation'. In many cases, this sort of obstructive behaviour on the part of decision makers only becomes noticeable much later on when other enterprises that have employed the technology have achieved competitive advantages. Enterprises that tolerate such behaviour run the risk of failing to devote adequate attention to strategic issues and suffering painful competitive disadvantages as a result.

Managers who are reluctant to make decisions and thus reject RFID technology on this superficial basis also spare themselves the subsequent steps in the process. For instance, introducing RFID requires the mutual agreement of all parties in the supply chain with regard to financing the antennas, readers, tags and software components of RFID systems. However, an enterprise might decide to wait and see whether another enterprise somewhere upstream in the chain is willing to commit to introducing RFID tags and bearing the associated costs. Then the 'wait-and-see' enterprise could use the tags in its own processes without having to pay for them. Naturally, this sort of attitude does not foster a spirit of trust and cooperation.

As can be seen from the description of the RFID hype and its causes, RFID technology must be regarded as a very diverse subject. The highly simplified information on this subject communicated in recent years has created a broad interest in the technology, but it has also created distorted perspectives. In summary, it can be said that RFID technology has the potential to make entirely new business processes possible, but just as in a football world cup where a young talent is prematurely elevated to the position of the saviour of a mediocre side and burns out under the burden of the responsibility, RFID concepts can suffer major harm if they are presented as patent remedies for poorly organized enterprise process structures. Enterprises that allow this to happen run the risk of sinking into the Gartner 'trough of disillusionment' due to their inability to fulfil inflated expectations. By contrast, realists will find the path to maturity with RFID.

2.1   Process Drivers

2.1.1   RFID as a Catalyst

The greatest change potential of RFID is generally expected to be found in logistics. There the main focus is on supply chain management (SCM), which means coordinating material and information flows across enterprise boundaries for the entire value process with the aim of achieving overall time and cost optimization [Scho1999]. The desired optimum can be arbitrarily complex, depending on the nature of the supply chain. Even the simple supply chain shown in Figure 2.5 yields significant challenges due to interaction effects.

Forrester studied the system dynamics of simple supply chains as early as 1958. He presented an explanation for the accumulation of excessive stock levels due to demand amplification in the reverse feedback path, which is called the 'Forrester effect' or more commonly the 'bullwhip effect'.[4] The bullwhip effect refers the fact that within a supply chain, the variation in demand amplitude increases disproportionately in the direction of the source. There are four principal causes for this increase in variability [Kel2004]:

- **Demand forecasts.** Supply chain participants forecast their future sales from previous key figures and add a safety margin to compensate for lead time. The retailer can keep the margin relatively small due to its proximity to the consumer, but this behaviour produces a cumulative effect in the subsequent feedback stages that produces large variations.

- **Shortage gaming.** If a shortage situation arises due to high market demand, the upstream stages will ration their products such that their customers, such as the retailer, receive products in proportion to their previous order quantities. The retailer will thus attempt to compensate for the feared reduction by increasing its current order quantity. This

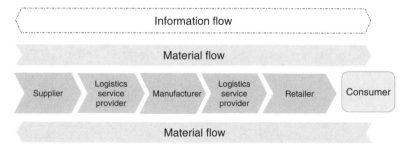

Figure 2.5   Model of a three-stage linear supply chain

---

[4] See Chapter 4 for a more detailed description.

can cause the supplier at the head of the chain to drastically misjudge the market situation due to lack of sufficient information.[5]

- **Order batching.** Customers attempt to batch their orders due to discount incentives and lower relative processing costs. This batching amplifies the bullwhip effect.

- **Price fluctuations.** If a product is subject to wide price fluctuations, customers are tempted to build up reserves when prices are low.

This analysis of the bullwhip effect leads to a clear realization that the enterprises in the supply chain can only reduce the effect by collaboration, or in other words by working together. 'Efficient consumer response' (ECR)[6] is a method for dealing with this situation [Vog2004]. Progressive enterprises have been devoting attention to this subject for more than a decade already. One of the main pillars of ECR is using a standardized method (EDI) to provide and exchange data, especially specific point-of-sale (POS) data, and conveying this data to the upstream links of the supply chain in near real time. RFID can support this process, and with its higher degree of automation and ability to identify individual items it can surpass existing barcode systems.

Providing POS data on the basis of item barcodes would unquestionably be possible even now. However, exchanging this sort of data is still hampered by resistance arising from business policies.

Although the bullwhip effect has been known and understood for nearly 50 years and concepts such as ECR are available, up to now it has not been possible to significantly constrain the effect. One reason for this can be seen from examining statistics on the extent of use of EDI and enterprise resource planning (ERP) systems.[7] There is still a large shortfall in the use of such systems, especially among relatively small enterprises. If RFID applications are to be used effectively, it is important that the data read using RFID are also processed properly in background systems, which means ERP systems, and that packing slips and similar items are transmitted by EDI. RFID cannot make any effective contribution in enterprises where these systems are totally absent or so fragmented that they cannot even provide consistent support for internal business processes.

Figure 2.6 shows the results of the 2005 European e-Business Report [ebus05] on EDI use. Only 37% of the enterprises in the food industry supply chain and, even more remarkably, only 21% of the enterprises in the garment industry supply chain make any use of EDI. Large enterprises (more than 250 employees) are already more advanced with 43%. By contrast, the result for small and medium-size enterprises (SMEs)

---

[5] This effect is very pronounced in the semiconductor industry, which regularly experiences allocation cycles.

[6] See Section 4.2.3.

[7] See Chapter 6.

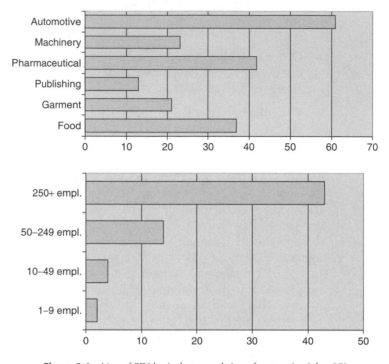

Figure 2.6   Use of EDI by industry and size of enterprise [ebus05]

is shocking. Only 14% of the enterprises in this category (50–250 employees) have implemented EDI methods.

These are the figures for the quantitative aspect of EDI use. If we examine the qualitative aspect, we see that essentially only two types of messages are exchanged: invoices and orders [Hors2005]. Exchanging information relevant to planning, such as daily sales data, is still far away.

This shows one of the essential problem areas in the implementation of 'enabling technologies' such as EDI: they must be seen as part of a whole consisting of complex business structures and processes. The status of internal IT integration of business processes is thus a good indicator of the maturity of an enterprise and its ability to integrate processes such as supply chain management that extend across enterprise boundaries.

A metric for this indicator is the use of ERP systems. Survey results for this are shown in Figure 2.7. Only somewhat less than 40% of the consumer goods industry (represented here by foods and garments) uses ERP systems. In terms of relative enterprise size, small enterprises occupy a clearly unfavourable position. Here again the SME sector scores distinctly worse, with only 33% versus 59% for large enterprises.

Even among the enterprises that do use ERP systems, not even half of SMEs use EDI. What is the reason for this reluctance? Impediments to the use of EDI were investigated in an ECR study conducted by the

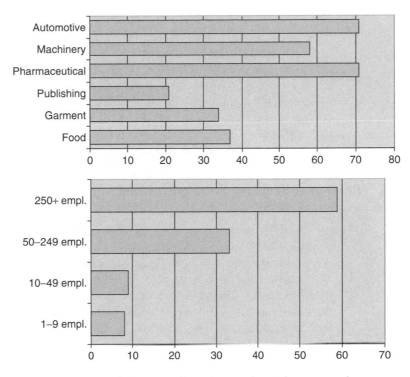

Figure 2.7 Use of ERP systems by industry and size of enterprise [ebus05]

Cologne-based EHI Retail Institute (Eurohandelsinstitut). The study noted as 'surprising' the finding that enterprises said that problems with data transmission and data processing (mentioned by 40% of the respondents) were still more significant than costs [Hors2005] – which is hardly surprising in light of the low incidence of use of ERP systems and the deficient quality of the systems. Quite often, several 'organically evolved' system domains (also called legacy systems) are present in enterprises. Such legacies often arise from company mergers or because individual departments choose different system providers without considering the need for integration. Integrating these internal systems is already a major challenge, and linking systems across enterprise boundaries is even more difficult, but end-to-end support for business processes is not possible without system integration.

Solving the technical problems is only one side of the issue. Another side is that a suboptimal state will prevail as long as decisions in the supply chain are predominantly made at the local or individual level with the aim of optimizing internal business processes, and under these conditions exchanging data by EDI is not by itself sufficient to generate integration along the supply chain.

What is important here is cooperation in a spirit of partnership and understanding that it is possible to achieve more with trust-based

cooperation and transparent planning. Only then is it possible to implement integration projects that span enterprise boundaries.

Due to interactions between process components, system components
and social components, a strategy aimed at maximizing collective benefits
(cooperation) is preferable to a strategy aimed at maximizing relative profits (competition)[Anan2005]. Here RFID technology stimulates process
change at the meta level. It serves as a common denominator or driver
for suppliers and customers to progress along the path of collaborative[8]
process adaptation, global system standardization and building trust.

2.1.2   RFID as a Basis for Future-Proofing

Even in a simple linear supply chain, planning uncertainty is higher with
decentralized optimization and decision structures. A lack of information
combined with small demand variations at the consumer end side cause
demand variations in the upstream stages to increase disproportionately.

Demand variations have also increased sharply in the industry and
trade sectors in the past five years, and they must be countered by
increased flexibility and shorter response times for planning of production
and sales volumes. The reasons for this can be briefly outlined as
follows [Strau2005]:

- increased internationalization of markets;

- reduced customer loyalty ('smart shoppers');

- shorter product life cycles;

- reduced predictability of competition strategies.

The process structure of a linear supply chain assumed in the past is
no longer adequate for modelling actual processes. Figure 2.8 shows a
supply network model, which more accurately reflects present conditions.
The web of relationships has become more complex, and simple bilateral
communication paths are increasingly being supplanted by multilateral
many-to-many structures.

Traditional EDI concepts are based on using a push process to transfer
records in specified formats based on defined responsibilities for retrieving
and providing data. In the specific case of a supplier notification,[9] the
supplier sends a defined message to the recipient of the goods when the
goods are shipped. This notification contains all relevant information. A
combination of bilateral communication agreements will quickly reach
its limits in a global network of supplier relationships with its language

---

[8] 'Collaboration' is often used to refer to cooperation across enterprise boundaries.
[9] An example of an EDI message format is EANCOM for the trade sector as defined by
GS1; there are also other formats for other sectors. They are described in detail in Chapter 4.

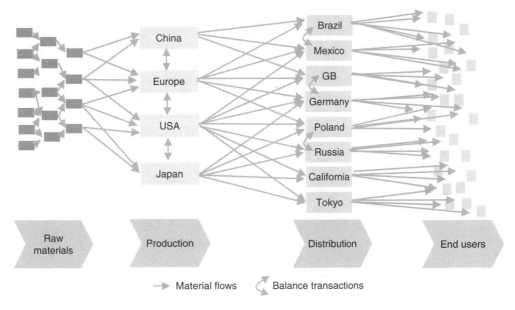

Material flows   Balance transactions

Figure 2.8 A complex supply network model

barriers, time offsets, wide variety of national requirements with regard to aspects such as customs designations, and so on.

Consequently, a properly functioning information platform must be constructed according to the pull principle, just like the Internet. The contents provided can be read or retrieved by persons or systems having suitable address information and access privileges. Here the responsibilities are shared by the information providers, which must ensure that the information is accurate and current, and the users, which must request the information on their own initiative. Nowadays we find it perfectly normal to navigate the Internet and use its hypertext structures to acquire relevant information. In exactly the same way, individually identifiable objects can be regarded as links to nodes in a network of objects. This is the underlying idea of the 'internet of things'.[10]

RFID is the only technology that can provide this individualization of objects and automatic data retrieval without any media break. It is thus overwhelmingly important to equip all objects in the material flow with RFID tags in order to fashion the supply chains of the future.

2.2 Security as a Driver

The attacks on 11 September 2001 marked a turning point in history with perceptible consequences for international trade. Since then, numerous measures for protection against international terrorism have been

---

[10] See Chapter 5.

implemented in the USA under the mantle of Homeland Security. Identification of persons and objects in international traffic is essential for
obtaining comprehensive knowledge with regard to goods importing and
traveller entry processes. The objective here is always to be one step
ahead, which means always having a decisive information advantage
regarding the location and status of identified objects and persons. RFID
is generally regarded as a support technology for this purpose.

One result of theses efforts is the introduction of an internationally
standardized electronic passport.[11] It contains an RFID tag that communicates via the air interface (radio signals). Naturally, this tag must be
a highly secure product because it contains more than just a number
or a small amount of data – it contains biometric information about the
passport holder and a security key. The chip in this tag is comparable
to a miniature PC and operating system with the performance level of a
Pentium 1 processor, and it executes cryptographic algorithms on a die
with an area of approximately 10 mm^2.

This sort of biometric passport has been issued in Germany since
November 2005. In this way a statutory provision based on stricter
security requirements fosters the use of RFID technology.

In the following, we examine other drivers that are not based exclusively on economic or commercial considerations, but instead serve to
implement legal and ethical constraints.

2.2.1   Transportation and Traffic Security

Along with electronic passports, there are also other approaches to
increased travel security. Particularly in the case of air travel, the airlines belonging to the International Air Transport Association (IATA)
are developing internationally standardized security processes. The main
consideration here is physical unity of passengers and their baggage in the
aircraft. If a passenger does not board an aircraft despite having checked
in, the passenger's checked baggage is removed from the airplane at
considerable effort. Increased numbers of aircraft passengers and new
aircraft designs such as the Airbus 380 lead to increased demands for
implementing innovative methods to ensure security.

Stimulated by the Homeland Security programme of the US government, experts have been working for several years already to develop
methods for comprehensive tracking of sea cargo containers. The objective is to transmit advance electronic information on container contents
and destinations to official bodies in the USA even before the containers
are loaded on board at the point of origin (such as the port of Rotterdam
or Hamburg) so that undesirable containers can be refused in advance

---

[11] Travel documents are standardized by the International Civil Aviation Organization
(ICAO). Electronic passports are governed by ICAO 9303-1 (CY 2004) and EU Regulation
2252/2004.

and arriving containers can be cleared quickly by simply reading out encrypted data. However, a prerequisite for reliable real-time inspection is the assurance that the contents of the containers cannot be manipulated while they are en route. In order to prevent this or allow it to be detected, in the future electronic seals (e-seals) based on RFID technology will be used. International operability of these concepts is ensured by provisions of the International Organization for Standardization (ISO).[12]

2.2.2 Supply Chain Security

Heightened requirements for tracking cargo containers arise initially from the desire to increase the transparency of the supply chain and obtain better control of goods flows, but increased transparency also helps develop solutions to requirements for security in the supply chain. RFID is a suitable technology for bringing a large number of threat scenarios under control. These requirements thus form an additional driver for RFID technology. Other requirements discussed below arise from grey-market issues, product protection, and consumer and patient protection.

Potential threats are present in supply chains, and the degree of the associated risk must be assessed according to the specific industry. Current discussions focus on two key issues:

- theft and shrinkage;
- grey markets and re-imports.

Theft and shrinkage

There are a variety of causes of disappearance of goods in the supply chain up to the point of sale (POS). The following causes are cited in statistics for inventory discrepancies [Hors2005]:

- Organizational deficiencies: 17%.
- Customer theft: 53%.
- Employee theft: 22%.
- Theft by suppliers and service employees: 8%.

The magnitude of the problem is clearly shown by inventory discrepancies of approximately €4 billion in Germany and €30 billion in Europe. Shoplifting alone causes damage to national economies amounting to approximately €230 million per year due to foregone tax revenues. On average, an inventory discrepancy of just 1.1% can be enough to deprive the affected enterprise of its entire profit, especially in the retail sector.

---

[12] See Chapter 5.

The highest losses due to inventory discrepancies occur in the retail sector. Detailed statistical data is available for this sector due to the magnitude of the losses. Similar statistics are not available for the industrial sector, but the problem is also present in other supply chains such as automotive and pharmaceutical.

With regard to theft, RFID can be used with alarm-generation functions to increase transparency in the supply chain or at the POS and thus make theft more risky for potential perpetrators. Electronic article surveillance (EAS) systems, which have been used in department stores for many years already, are equivalent to 1-bit ('paid/not paid') transponders.

Grey markets and re-imports

Grey markets and re-imports arise from price or tax differences between individual countries, which are exploited by clever merchants to offer products at lower prices or increase their profit margins. These practices are especially common inside the EU, where the member states pursue national policies while at the same time encouraging free trade. This leads to differentials that are exploited by legitimate or illegitimate market players. For example, liberalization of the automobile market in the EU in 2005 led to the establishment of new, legitimate distribution structures and made re-imports attractive. Grey-market activities that are damaging from the perspective of national economies are found in the cigarette and computer areas. Grey markets and re-import activities that primarily cause damage to business economies are found in the luxury consumer goods area, such as perfumes and other high-ticket commercial items. Here the primary concern is purchaser disappointment with the brand despite the fact that the manufacturer is not to blame for the counterfeiting. The brand image suffers as a result.

RFID can provide comprehensive goods traceability and thereby make a major contribution to hindering or identifying grey-market and counterfeit products.

2.2.3   Consumer Protection and Patient Protection

Brand fraud and product piracy are constantly expanding in volume, technical sophistication and scope. In addition to causing economic damage to the affected enterprises, this expansion damages the national economies of industrialized countries. As the activities of product counterfeiters are no longer limited to music, software and luxury goods, but instead are increasingly focused on pharmaceutical products, replacement parts and commodities, they can also create real hazards for consumers [Staake2005].

The OECD estimates the size of the counterfeit market to be 5 to 7% of the total world market. It is thus an activity that must be taken

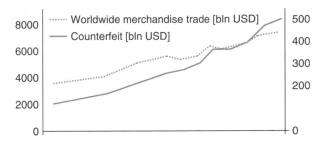

Figure 2.9 Market evolution of counterfeit products compared with the worldwide market [Staake2005]

seriously, and it has a professional, industrial character. Figure 2.9 shows the evolution of the illegitimate market for counterfeit products compared with the evolution of the worldwide market.

If products are counterfeited in order to enable consumer groups with relatively low spending power to acquire status symbols by means of a brand image, this diminishes the value of the brand for its owner (a qualitative, psychological loss) and thus reduces the owner's revenue (a quantitative, monetary loss). The national economies of the countries where the brand owner is legally registered suffer lost tax revenue and reduced willingness to undertake innovation, so they risk losing jobs in their internal markets. The European Economic Community has responded to this by means of the European Commission, which has presented a 'package of measures to strengthen protection for the EU and its citizens against counterfeiting and piracy'.[13]

However, a considerably greater risk arises from counterfeiting of foods, pharmaceutical products and replacement parts for motor vehicles and aircraft. The resulting hazards lead to a debate that is not only driven by economic interests, but also distinguished by ethical considerations.

At least since the US Food and Drug Administration (FDA) issued a recommendation for using RFID to combat counterfeiting of pharmaceutical products,[14] active efforts are being made to curtail counterfeits of this sort, which have undeniable consequences for the health of the general population. Use of RFID is explicitly recommended by the FDA. The aim is to achieve complete visibility of the product life cycle from the raw materials to the consumer ('product pedigree'). In the opinion of experts, the FDA's recommendation to use RFID to help combat counterfeiting of pharmaceutical products is currently the most important driver for the use of RFID in the pharmaceutical sector. Although compliance with the FDA recommendation is voluntary, the agency's recommendation appears to be unambiguous and unavoidable [Ber2005].

---

[13] European Commission, IP/05/1247 of 11 November 2005.
[14] *Combating Counterfeiting Drugs: A Report of the Food and Drug Administration*, FDA, February 2004.

Traceability of 'food, feed, food-producing animals and any other sub-stances' is prescribed by a directive of the European Union since 1 January 2005. According to the directive, participants in the supply chain must establish 'systems and procedures which allow for this information to be made available to the competent authorities on demand'. In contrast to the recommendations of the FDA, no specific technologies are mentioned in the directive, but the necessity for unambiguous identification of goods in the supply chain automatically leads to an intense interest in RFID technology.

Due to potential risks to aviation safety resulting from counterfeit replacement parts with reduced capability, aircraft manufacturers Airbus and Boeing have joined forces to formulate requirements applicable to suppliers. For this purpose, both companies have conducted long-term tests to investigate RFID applications in aircraft and have subjected these to the stringent approval procedures of the US Federal Aviation Administration, which is responsible for aviation safety.

In addition to assured original quality, quality assurance along the entire supply chain is increasingly drawing the attention of consumer protection organizations. This includes compliance with specified trans-portation conditions:

- maintaining the cold chain;

- detecting harmful vibrations and shocks;

- providing evidence of tampering (contamination) for the purpose of extortion.

Complete documentation of the handling of each individual product is necessary. RFID tags that can record temperature histories are available, and they are already being used.

2.3 Mandates, Mentors and Sponsors

Only a few technologies have experienced a level of international interest as early as the start of the maturity process that compares with the interest in RFID technology. RFID is classified as a horizontal technology, which puts it in the same league as nanotechnology, biotechnology, information technology and microelectronics. Strictly speaking, RFID is only a subset of information technology or microelectronics, but as already mentioned in the previous description of the RFID hype, RFID is overloaded with expectations regarding its potential for independently affecting our ability to automate processes. It is more realistic to regard it as an enabling technology that can only realize its potential in combination with other technologies and organizational measures.

In contrast to the above-mentioned horizontal technologies, which every enterprise can use independently as a basis for new products and innovations, RFID is dependent on collaborative applications that span enterprise boundaries – at least for wide-scale use of the technology. RFID technology has the potential to eliminate media discontinuities in data communication, and above all to foster integration of different types of enterprises in a common supply chain.

Users in the trade and scientific sectors in particular recognized this at a very early stage, and they launched a concentrated initiative (EPCglobal) in the early days of the technology to define specifications for using RFID and collaborate on implementing them in a staged process. If individual enterprises had taken the lead with proprietary solutions, this would have quickly led to a divergent system landscape that would hamper consistent integration of supply chains.

Based on this insight, the Auto-ID Center was founded in 1999 at the Massachusetts Institute of Technology (MIT) in Cambridge (USA) with the objective of developing requirement profiles and standards based on future expectations for a 'global network of physical objects', or in other words, global supply chains.[15]

This initially American initiative was founded by MIT in cooperation with the nonprofit Universal Code Council (UCC)[16] and two commercial enterprises: Procter & Gamble (P&G) and Gillette. Its activities grew quickly, and it attracted new members in Europe that enabled it to address international demands, such as Tesco and Marks & Spencer in the UK, Ahold in the Netherlands, and the Metro Group in Germany.

A comprehensive restructuring of the membership occurred in 2003. The umbrella organization GS1 International[17] arose from the merger of former barcode organizations UCC and EAN International. The activities of the Auto-ID Center were transferred to the newly founded EPCglobal Inc. and the Auto-ID Labs.[18] These two organizations became subsidiaries of GS1 International. At this time, the EPCglobal initiative already counted more than 100 enterprises worldwide among its members. Many leading figures of its member enterprises are members of the board of directors of GS1 (see Section 5.2 for more details).

There are Auto-ID labs associated with MIT (USA), Cambridge University (UK), and the universities of Adelaide (Australia), St Gallen (Switzerland), Mujiro (South Korea), Fujisawa (Japan) and Shanghai (China).

Another factor that increased the significance of EPCglobal, beside internationalization, was expanding its membership to include nonindustrial institutions such as government bodies (e.g. the FDA) and the US Department of Defense (DoD). A total of approximately US$70 million in

---

[15] www.autoidlabs.org/aboutthelabs.html.
[16] The US counterpart of EAN.
[17] GS1 stands for Global Standard 1 (see www.gs1.org).
[18] See www.epcglobalinc.org and www.autoidlabs.org.

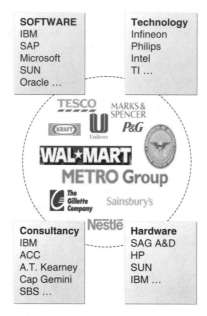

Figure 2.10 Enterprises that have participated in the concept generation activities of EPCglobal and the Auto-ID Center

sponsor funding for the activities of the Auto-IC Center was raised as early as the autumn of 2003. The member enterprises, some of which are listed in Figure 2.10, represent more than US$500 billion in annual turnover. Given this solid financial base, a professional organizational structure was implemented with dedicated working groups to generate specifications for hardware (Hardware Action Group), software (Software Action Group) and processes (Business Action Group). The working groups devised and presented the basic concept for the 'internet of things'.

From the very start, a key aspect of the concept was the Electronic Product Code (EPC), which can be regarded as an extension of the Universal Product Code (UPC) developed by UCC and used in North America and the European Article Number (EAN), both of which are commonly used in bar codes. This clearly shows the predominant influence of the consumer goods industry on this initiative, which is also reflected in the composition of the membership of EPCglobal. EPC is a universal code with a length of 96 bits, which is sufficient for internationally unique identification of practically all objects or items of significance to trade and industrial processes.[19]

---

[19] Numerical example: a 96-bit code can assume $2^{96} = 79\,228\,162\,514\,264\,337\,593\,543\,950\,336$ (7.92×10^{28}) distinct values. The estimated number of individual products in international supply chains is approximately 500 billion (500×10^9) per year. If all objects were assigned unique codes starting in 2005, it would take 100 years to use up 5×10^{13} codes.

An EPC by itself does not provide any information about the associated object. Access to a database in an IT system is necessary to obtain a description of the object and its attributes. The objects are listed in the database by their EPC numbers. Consequently, the EPC concept requires maintaining extensive IT infrastructures with two essential features:

1. Integration of the reader periphery into networks that span enterprise boundaries. To support this, EPCglobal has developed the Savant concept,[20] which governs standardized access to the IT systems of participating enterprises via EPC. The corresponding software is also called 'edgeware'.

2. Networking of the IT systems of the participating enterprises with the Internet via web servers and the still to be developed 'internet of things'.

These challenges stimulated many software and IT enterprises to join EPCglobal and participate in the initial development of these concepts (see Figure 2.10).

In addition to direct contributions for membership in EPCglobal or the Auto-ID Center, sums in the millions have been expended on EPC marketing activities of the participating enterprises over the years, especially among blue-chip companies such as IBM, Accenture, the Metro Group and Wal-Mart. As a result of numerous conferences and studies, the term 'RFID' has become closely associated with 'EPC', and there is already a profitable market in the quaternary business sector.

The most striking aspect of all these activities is the diligence with which the members pursue the EPC theme. This has also contributed to accelerated integration of the economic spheres of the USA, Europe and the Asia/Pacific region. The catalytic force of RFID is thus effective even at this level.

Beside the powerful GS1/EPCglobal consortium, which created the theoretical framework, other initiatives for implementation of the concept arose starting as early as 2003. The essential milestones are shown on a timeline in Figure 2.11. This process is dominated by the 'mandates' of Wal-Mart and the Metro Group. They specify staged schedules for using RFID tags with mandatory deadlines, and the suppliers of these retail enterprises must comply with these requirements. This applies in particular to RFID tags on pallets, cases and individual items. These mandated requirements create a strong pressure to introduce RFID. The rationale behind this is naturally that the retailers can obtain the greatest benefits from items that have RFID tags.

Wal-Mart holds the top position in food retailing with a turnover of nearly €230 billion (2005). The second position is held by French retail

---

[20] Savant means 'learned person'.

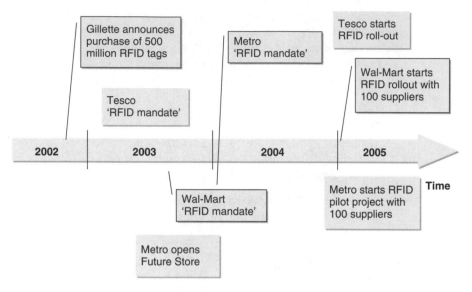

Figure 2.11   Early milestones in the implementation of RFID pilot projects

giant Carrefour with approximately €73 billion, followed by the Metro Group with more than €56 billion. The expression 'gorilla in the supply chain' [Gaja2002] used in the USA to refer to dominant occupants of a link in the value chain has applied to Wal-Mart for a long time already, and at least in Europe the term can justifiably be applied to Metro as well.

With its 'Future Store' concept,[21] Metro has even managed to develop an international thematic leadership position. The Future Store is a supermarket in Rheinberg near Düsseldorf, Germany, in which RFID methods are tested in practical use. It is complemented by the RFID Innovation Center in Neuß, which is a demonstration centre for RFID applications in the retail sector that Metro operates in conjunction with its system partners, and which was one of the highlights of the RFID topics at the CeBIT 2006 fair. All trade sector applications and process steps related to RFID are demonstrated and tested in the Future Store. To mention a few examples in the retail area, there is an intelligent changing room that suggests suitable accessories for a suit, a refrigerator that knows what is inside it, and a checkout (POS) that generates sales slips automatically. In the logistics area, among other things there is a demonstration of how pallets can be read and individual clothing items on hangers can be sorted automatically according to different destinations.

However, the hype rears its head here as well, since the tags shown on the individual items in the refrigerator are not yet available in reality. According to Metro, this sort of tagging will come in five to ten years.

---

[21] See www.future-store.org.

Retail trade is a very strong theme in the RFID context because it is where potential uses can be identified most easily, as it involves global processes that span enterprise boundaries and can profit from RFID, and many respected bodies and organizations pursue internationally oriented communication campaigns in cooperation with GS1, EPCglobal and Auto-ID Labs. However, it is nevertheless necessary to avoid creating the impression that RFID is only significant in the consumer goods chain. Other sectors such as the automotive, aviation, defence, high-tech and chemical/pharmaceutical sectors exhibit comparable groupings. In this regard, Boeing and Airbus have a very public orientation. They are on their way to allowing RFID tags to be attached to their aircraft, and they are working very closely with their suppliers to achieve this.

In summary, we can say that RFID can develop into a very powerful lever for innovation and integration in many areas of industry, but enterprises must look for the right places to apply this leverage. This should make it possible to eliminate impediments to the fastest possible implementation of RFID systems.

2.4 Technology Drivers

2.4.1 Ubiquitous Informatization and Intelligent Objects

Technology convergence, which means the mutual convergence of information and telecommunication technologies with media content, is a current phenomenon in our information society. The evolutionary merging process described by this term is already eroding the hard distinction between technology platforms and associated applications, with a distinction between digital resources arising in its place. The boundaries between the physical and virtual worlds are becoming blurred. Concrete aspects of a new IT paradigm have recently begun to emerge gradually along these nebulous boundaries (Figure 2.12). Following the succession of mainframe computers, client/server systems and mobile devices, individual objects are now poised to become bearers of information and 'intelligence'.

A combination of increasing miniaturization and decreasing prices facilitates a massive rise in the use of memory, processors, sensors and actuators in all aspects of private and public life. The number of processors in an upscale car is already in the two-digit range. Our environment is coming to be characterized by 'ubiquitous informatization'. This development was first described systematically in the 1990s by Mark Weiser [Weis1991], among others. However, 'ubiquitous computing' is not the only generic term that is used to describe this phenomenon. 'Pervasive computing' and 'sensor networks', among other terms, are occasionally used as synonyms for 'ubiquitous computing'. Figure 2.13 represents an attempt to differentiate these terms and illustrate their meanings.

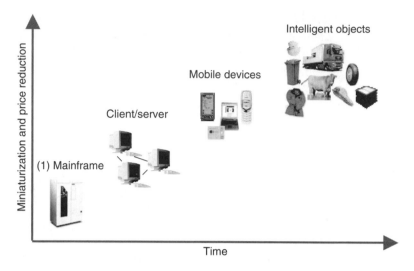

Figure 2.12 Evolution of IT paradigms [Flei2005]

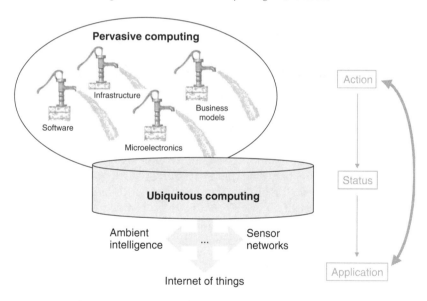

Figure 2.13 A suggested demarcation of the new IT concepts

'Ubiquitous' means 'omnipresent', while 'pervasive' means 'widespread'. One way to distinguish between these two concepts arises from the idea that 'pervasive' characterizes a process – in this case progressive networking of the environment on the basis of existing technologies and innovative business models – while 'ubiquitous' describes a status.

According to this distinction, the result of pervasive computing activities is a certain degree of ubiquitous informatization, similar to the level of water in a basin this is being filled in a figurative sense by pervasive computing. More concrete, application-specific terms such as 'ambient intelligence' (human/environment relationship), 'internet of things' and 'sensor networks' (decentralized/mobile infrastructures) can generally be regarded as applications of ubiquitous technologies, methods and processes. Pervasive computing and ubiquitous applications interact with each other.

Despite this attempt at demarcation, the meanings of the terms are not sharply defined, and one or the other of them will probably come to prevail in the course of further development of the underlying concepts. In this book, 'ubiquitous computing' is used to describe the phenomenon from a general perspective, and 'internet of things' is used for aspects related to RFID.

3

The RFID Market

Unlike most portrayals of the RFID market, the analysis presented in this chapter does not reveal any splendid growth rates. We leave that to the relevant market studies. Our objective here is to analyse the structure of the market and estimate the changes that can be expected. There is no doubt that the market will be large (see [Flei2006]). However, it is not possible to give a definite answer to the question of how fast it will develop. Incorrect forecasts have often been made in recent years. When and how enterprises address the subject of RFID is a matter of individual choice, and it requires a careful analysis of the existing situation. An understanding of market and application structures provides meaningful support for decisions regarding formulation of proposals on the part of producers and planning for deployment of technologies in user projects. In this chapter, we differentiate between two perspectives on the RFID market: that of the users and that of the producers.

3.1 The RFID Value Chain

We start by analysing the RFID value chain. Figure 3.1 shows the basic components in context. Here we do not devote any attention to production of the integrated circuits (ICs), which in this case are RFID chips, but instead focus on the subsequent stages: production of the inlays (chip plus antenna) and tags (inlays integrated into housings, smart cards and the like), and of course production of reader software, and finally system integration.

The RFID market is still structured such that enterprises can focus on a single element of the value chain. Here we can start with tag production. The basis for this is an IC, which is a silicon-based semiconductor device. Semiconductors are produced in special fabrication units in the form of wafers. A wafer is a disc that can contain around 60 000 ICs, depending on the size of the IC (up to 1 mm^2 for RFID applications) and the diameter

RFID for the Optimization of Business Processes   Wolf-Ruediger Hansen and Frank Gillert
© 2008 John Wiley & Sons, Ltd

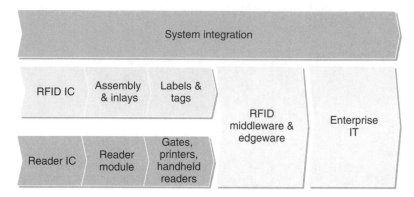

Figure 3.1 RFID value structures

of the wafer (8-inch wafers are currently standard, but 12-inch wafers will be used in the future).

ICs are usually made from silicon. However, polymer chips ('polytronics') will also be available in the future, and they will be especially important for integration into packaging. Polymer chips are also semiconductors, but they use materials other than silicon (see Section 4.4.3).

The potential production volume of a modern fabrication facility is around 10 000 wafers per week, so it is possible to produce 600 million ICs within a week. A fabrication facility can thus produce enough ICs in a few weeks to meet a future demand for several billion ICs, even without considering the possibility of further miniaturization.

The next stage is assembly, which is where the RFID inlays are produced. In this process, the individual ICs are separated from the wafer, an antenna is attached to the IC, and the ICs are attached to a carrier (such as a film). The antennas are usually produced by etching. Antenna production and assembly do not necessarily have to be performed by the same organization, but the trend is in this direction. The RFID inlay already has the full functionality of the later RFID tag.

The tags are produced in the next stage. The advantage of tags is that they are sufficiently robust for the intended use, they have surfaces that can be printed, and they can be attached to objects. Tag producers usually work closely with tag users and supply the components necessary for using RFID tags in the subsequent overall system.

The market on the reader side is similar. Semiconductor manufacturers supply reader ICs, which are highly integrated circuits for analogue to digital conversion of the received signals. The next stage consists of producing reader modules, which differ in terms of their intended working range and operating frequency. They form the basis for development of the final stages of the system, such as antenna gates (gate solutions), merchant devices (which combine an antenna, reader and display in a single unit), and printers with embedded RFID components. The same types of modules can be built into merchant devices and printers. These are the links in the value chain of the hardware components of an RFID system.

Readers are usually connected to a server hosting the RFID middleware. This is the first processing level for the data acquired by the RFID devices. The software components dedicated to the readers are also called edgeware. The edgeware provides the communication between the hardware and the middleware. There are two types of middleware here: the previously mentioned RFID middleware communicates with the IT system via the enterprise application integration (EAI) middleware, which in turn supplies data to the higher-level ERP systems. These layers are shown in simplified form in Figure 3.2.

As usual with new technologies and concepts, it will take a while for standardized terminology to develop and become established. There are thus occasionally differing opinions regarding the terms 'RFID middleware' and 'edgeware' that we have chosen to use here. However, 'RFID middleware' is at least widely accepted by analysts and authors of market studies. The term 'edgeware' is promoted vigorously by a few IT suppliers. The organization of the IT architecture is described in more detail in Chapter 6.

The use of RFID systems, and in particular implementation of the EPCglobal concept, leads to an enormous increase in the amount of data to be processed and the severity of the real-time requirements imposed on the systems. This is one of the essential reasons for the emergence of new RFID middleware, which filters and preprocesses the data that is read.

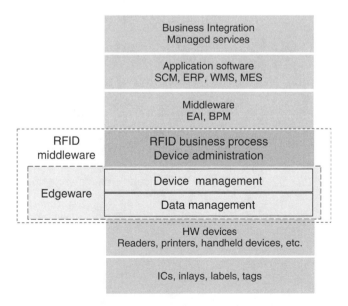

Figure 3.2 Simplified RFID layer model

As with the launch of PCs in the early 1980s, there are many suppliers of middleware, and the number is constantly growing. This market segment is thus very confused, and users must carefully examine the available products. A single dominant producer with a structure-defining role, such as has been assumed by Microsoft with the widespread use of PCs, is not (or not yet) visible here. This is in part because newly established middleware producers do not yet exhibit any exclusive features, and in part because well-known major suppliers are attempting to use their market strength to dominate this sector as well. Nevertheless, the RFID market is fostering a general paradigm shift towards decentralized IT infrastructures.

The stage for this revolution is the 'middleware platforms', which are supposed to make processes much cheaper, faster and more flexible and are roughly comparable to the shift from mainframe computers to PCs. The question of who will set the new standard – SAP with NetWeaver, IBM with WebSphere, Microsoft with .NET, or Oracle with Fusion – is still open.
Excerpt (translated) from 'Machtkampf um den nächsten IT Standard', *Süddeutsche Zeitung*, 20 October 2005

Due to the anticipated strategic importance of RFID middleware, existing suppliers of EAI middleware are expanding their product lines accordingly, and it can be assumed that this will lead to the emergence of a comprehensive, coherent software layer that fulfils both EAI and RFID requirements and can be distributed over various server levels (Figure 3.3).

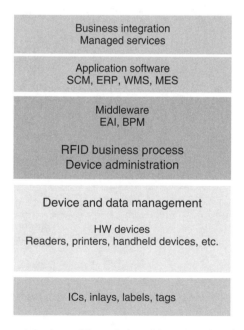

Figure 3.3 A possible evolution of the RFID value chain

RFID business logic and device administration will be merged into the middleware. Device administration includes in particular administering thousands of RFID readers that will be used in the future logistic networks of enterprises. The edgeware will increasingly be embedded in the devices or hardware infrastructure components, which means readers, network hubs and routers, programmable logic controllers (PLCs), and industrial PCs. The proportion of software embedded in chips will increase as well. Enterprises that already wish to start using RFID technology are thus confronted with the task of upgrading their equipment with each new generation. This iterative process also lies behind the fact that the necessary standards will only become available in stages. The current Generation 2 EPC standard is also only one step in the process, and it will be followed by several others. Nevertheless, we believe that these circumstances are typical for the launch of a new technology. Despite this situation, progressive enterprises will be able to extract short-term benefits from using RFID technology, and above all they will acquire experience in the early phase of the market that will pay off in the form of market advantages later on.

3.2 Trends in the RFID Market

As already mentioned, the RFID market is in a state of flux. This raises the issue of consolidation. In discussions with market players that could potentially be partners, one often encounters a tendency to set up barriers in situations where cooperation would be more reasonable. For instance, a tag producer may regard a system integrator as a competitor even though the two parties have essentially complementary roles in RFID systems. There are two possible causes for this situation:

- The roles of potential suppliers are still unclear in the present market environment. This uncertainty leads to a tactical endeavour to assume overall responsibility or the prime contractor role for projects, even when this does not correspond to the core competences of the enterprise or its provisional strategic positioning in the value chain.

- Customers want to obtain everything from a single source and cannot assess the risks, so RFID suppliers assume opportunistic roles that they later cannot fulfil properly.

Such problematic situations often occur when enterprises have made premature investments, such as in RFID hardware, without having an overall system concept.

Figure 3.4 shows the current situation from the perspective of suppliers of enterprise resource planning (ERP) systems, which are the central organs of the digital nervous system of an enterprise. It is thus understandable

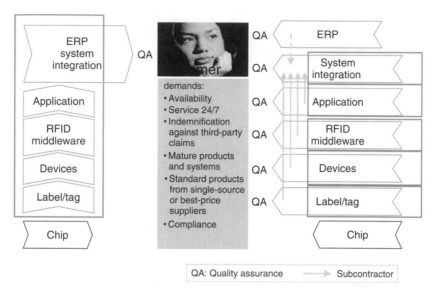

Figure 3.4 Market consolidation and assumption of responsibility

that users often approach ERP suppliers with the expectation that they will provide RFID solutions that match the IT structure of their ERP systems. In such cases, the ERP specialist is confronted with the challenge of structuring a system and assuming responsibility for overall quality and quality assurance (QA).

Customer requirements in such cases can extend beyond the usual standard of availability (24 hours a day, 7 days a week, etc.). For instance, assuming liability for patent claims brought by third parties, including as yet unknown claims, is often demanded. This is understandable from the customer perspective, but it creates major obstacles during the start-up phase of RFID projects. Here the ERP supplier must carefully consider how far its customer service should extend.

The situation and role distribution shown on the right-hand side of Figure 3.4 will materialize in the future.

In a study on logistics trends in industry and trade carried out in 2004 [tenHo04], the Fraunhofer Institute for Material Flow and Logistics (IML) (Dortmund) conducted a nationwide German survey of the expectations of the trade and industry sectors with regard to RFID. Survey participants were also asked about obstacles to the use of RFID. A similar survey was conducted in 2005 by Bremen-based Bundesvereinigung Logistik eV (BVL) [bvl2005]. Comparison of results of the two studies yields some interesting signs of increased market maturity between 2004 and 2005, which is also reflected in user attitudes. This comparison is presented in Figure 3.5.

The data from the BVL study reflect information from the producer segment, while the data from the IML study reflect information from the

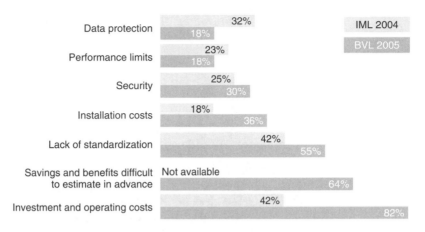

Figure 3.5 Obstacles to use of RFID (comparative figures for 2004 and 2005) (based on [tenHo04] and [bvl2005])

consumer and capital goods segments. The results are not based on fully comparable survey methods, but they allow conclusions to be drawn regarding trends. The IML survey used a six-level rating scale ranging from 1 (very significant) to 6 (not significant), while the BVL survey was based on a three-level scale. Consequently, lower percentages in the BVL results relative to the IML results can be regarded as significant because the BVL study combined two rating levels of the IML study into a single level. If the IML rating levels were merged, the difference between the two values would be even greater. If the BVL figure is higher than the IML figure, there was no change in the rating of the obstacle or the change was not significant.

In concrete terms, the obstacles to the use of RFID can thus be identified as follows:

- *Data protection – strongly decreasing significance.* Increased efforts to resolve this issue on the part of users and suppliers, as well as public institutions, have essentially removed doubts in this area (see Chapter 8).

- *Performance limits – strongly decreasing significance.* Developments in recent years and the standardization efforts of the EPCglobal network have resulted in performance improvements (essentially improved operating range by using UHF).

- *Security – unchanged or slightly increasing significance.* System security is tending to increase in significance, which reflects integration into processes and IT infrastructures. Security considerations become prominent when pilot systems are no longer tested offline.

- *Installation costs – slightly increasing significance.* The significance of installation costs is increasing, which is sign that enterprises are actually engaged in integrating RFID systems.

- *Lack of standardization – unchanged or slightly increasing significance.* Standardization is still an area of concern among enterprises.

- *Potential for cost savings and benefits difficult to judge in advance – very significant.* This wording was not included in the IML survey, perhaps because the initially propagated and accepted generic potentials were only studied as part of an implementation analysis. From this perspective, it can also be regarded as a positive signal in the sense that enterprises are addressing the issue intensively and professionally.

- *Investment and operating costs – strongly decreasing significance.* This point underpins what has already been said. If this issue increases in significance with decreasing component costs, this may arise from specific analysis of the relationship between RFID system integration costs and total cost of ownership (TCO).

3.3   Application-Specific Trends in the RFID Market

How can a supplier of RFID products optimize its activities in the current environment? Which application structures must users take into account in order to obtain realistic project plans?

The previously described obstacles arise primarily from two market barriers: a structure barrier and a cost barrier. The quadrant chart shown in Figure 3.6 is intended to illustrate how these barriers work and how they arise from various types of business processes.

In the vertical direction, the business processes shown on the chart are divided into closed-loop and open-loop types. 'Closed loop' designates a closed system in which the responsibility for all material, information and financial flows is internal to the enterprise. Fleisch uses the term 'unitary balance boundary' in this connection [Flei2005]. By contrast, 'open loop' designates a process that extends across two or more enterprises. Agreement between these enterprises is thus necessary in order to optimize the entire process, distribute the costs and allocate appropriate benefits to all participants. This is the classic case with a supply chain in the consumer goods sector.

The form of use of RFID tags is shown on the horizontal axis of Figure 3.6. 'Reusable' means that the tag remains attached to the product for its entire lifetime – such as a tag attached to a container, tool or library book. By contrast, tags on retail items are 'disposable' because they remain attached to the product as far as the consumer and pass through the logistics process only once. In the case of reusable tags (such as tags on returnable containers), the tag costs can be apportioned over their frequency of use and are thus significantly lower per pass through the process than the acquisition costs of the tags.

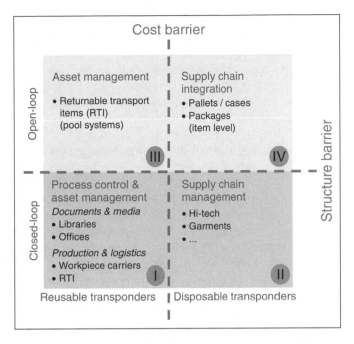

Figure 3.6 Market structure and application areas for RFID

In the case of disposable tags, the costs can be weighed against the benefits for the entire process chain. This often depends on the extent to which the tags tend to foster integration of the chain, which makes this a nontrivial case of cost/benefit analysis. As we already saw in Section 3.2, here the potential for cost savings and benefits is classified as difficult to estimate. A detailed analysis of the enterprise-spanning process chain is necessary for this analysis.

The coordinate scheme shown in Figure 3.6 yields four quadrants, with each pair divided vertically by a cost barrier and horizontally by a structure barrier. The tags to the left of the cost barrier are reused, so tag costs on this side approach zero due to repeated use. One-time use is necessary to the right of the barrier, which means the cost is incurred anew for each pass through the process.

Applications inside a single enterprise are shown below the structure barrier. This allows simple cost/benefit analyses and direct decision-making bodies. The processes above the structure barrier always involve two or more enterprises. This complicates cost/benefit analyses and decision processes. In addition, the strategic attitude of many enterprises, which are reluctant to create too much transparency with respect to their trading partners, acts as a hindrance here. For example, it is still difficult for suppliers in the trade sector to obtain sales statistics from retailers.

Finally, we see that the barriers decline over time: the cost barrier[1] declines with decreasing transponder prices, while the structure barrier declines with increasing standardization.

An awareness of these structures makes it easier for suppliers and users to get their bearings when planning projects and positioning their goods and services. The contents of the individual quadrants of Figure 3.6 are explained below using specific examples.

- **Process control and asset management.** Applications in this area are already in widespread use in production environments. Some libraries are increasingly utilizing RFID technology. A wide variety of applications are also available for document management. This is in part due to the combination of readily estimated benefits arising from simple business processes and transponder use extending over many decades. In the production control and logistics areas, there are also many years of experience with actual products and systems, such as for controlling workpiece carriers. However, returnable transport items (RTI) in the intralogistics area also represent an economically viable application area (see Chapter 4).

- **Supply chain management.** Applications in this area are already in common use in the garment industry because it is vertically structured. The enterprises are distributors with their own retail outlets. They make their own goods or have exclusive house brands made for them. However, there are also two other significant essential drivers here:

  - Garments have very complex logistics due to a high level of diversity (sizes and colours) and seasonal effects (very narrow sales window). This means that items that are not available on the sales floor at the right time can easily become unsaleable. RFID yields large potential benefits here due to increased precision in logistics processes.

  - Garments are usually produced in foreign countries (such as Tunisia or China). To optimize production control, vertically structured enterprises employ data transmission systems (such as EDI) and flexible product provisioning systems (such as tags). At regional sites, blank hanger tags are updated with data extending as far as special discount prices and then delivered to sales sites. The data used in the RFID tags can easily be stored in the tags.

- **Asset management.** Asset management in the open-loop area encompasses in particular returnable transport packaging systems. These

---

[1] Besides transponder reusability, value aggregation relative to the transponder price is also a significant factor, such as with entire pallets fitted with pallet transponders (Metro and Wal-Mart applications).

business processes are discussed in Chapter 4. In contrast to the closed-loop situation, realization of the indicated benefits of RFID cannot be accelerated due to the structure barrier. This is due in part to the need to generate the necessary standards, and in part to the fact that equipping existing pools of containers (numbering in the millions) with RFID tags poses a major logistics challenge. Nevertheless, an upturn can be expected here in the near future because provisions for using RFID are included in the planned replacement of existing pools of containers.

- **Supply chain integration.** Supply chain management is an application area with international dimensions, and it accordingly receives the largest amount of attention. It is a focus of the activities of the EPCglobal consortium, among other organizations. At EPCglobal, supply chain integration is one of the basic elements of the concepts for collaboration between trading partners – in particular efficient consumer response (ECR) and collaborative planning, forecasting and replenishment (CPFR). RFID supports comprehensive data integration in the 'real-time enterprise' context. The hurdles standing in the way of implementing these concepts are the extremely high costs and severe structural obstacles. Consequently, implementation is proceeding in stages, for example in the trade sector by using RFID tags first with pallets and later on with cases. This initially defuses the two factors that create barriers. Actual use occurs initially at the retail end, with supply chain integration being limited to transmission of despatch notifications by producers. The cost/benefit ratio can be calculated due to value aggregation at the load unit level.

As already described in Chapter 2 with regard to the technology management perspective, the abbreviation 'RFID' is becoming a synonym for a wide variety of applications. The media hype around the use of RFID in the supply chain (quadrant IV) overshadows its economic successes in the asset management and process control areas (quadrant I). RFID is already a basic technology in the latter area due to the relatively low influence of the cost and structure barriers. However, this does not mean that suppliers or users should limit their efforts to 'cherry picking' in the first quadrant. The economic success of small-scale RFID projects not only acts as a motor for evolution of the technology in terms of functionality and cost, but also gives an additional boost to the standardization process.

EPCglobal represents the essential standardization process that is necessary to release the extraordinary potential of applications in the fourth quadrant and thus enable suppliers to participate in the resulting market. The quadrant chart shown in Figure 3.6 can also help RFID users decide which entry scenario best matches their situation and which steps can be planned to ramp up their activities. The cost and structure barriers for subsequent applications can be eliminated by the combination of utilization of RFID technology and rapid progress of standardization. This will enable RFID to evolve further towards the status of a basic technology for supply chain applications.

Suppliers can use the quadrant chart to help them structure their range of goods and services and plan their investments rationally. A large number of market players have withdrawn from the RFID business in the past five years, perhaps because they did not take a sufficiently differentiated view of the market structures and selected the wrong strategy. This not only damages the individual enterprises, but also casts the wrong light on the RFID market, which in turn slows down the development process. The enormous market figures presented in the relevant studies are unquestionably realistic, but they will not materialize automatically. Instead, they can only be achieved by determined effort on the part of all parties concerned.

4

Business Process Structures

As already described in previous chapters, there are many different application areas for RFID technology. Here we concentrate on business process structures that have extensive potential for initiating changes from the logistics perspective. The term 'logistics' has an equally extensive range of use, so we must first define the scope of our analysis (Figure 4.1).

At present, most attention is being devoted to potential automation of the physical flow of goods between producers of consumer goods and retailers. Due to feasibility considerations, current system concepts are limited to using RFID to label pallets and cases. The individual processes supported here are assigning the correct EPC code to the shipment during outgoing goods processing by the producer of the consumer goods and triggering an advance information flow in the form of a despatch advice message (DESADV).[1] The retailer can then receive the goods using automated methods and compare the data with the despatch advice. This limited use of RFID creates an unbalanced distribution of benefits for the involved parties, since the highly automated processes of the consumer goods producer, in conjunction with the logistics strategies of the retailer (self-service and cross-docking), do not yield any benefits at the pallet level and only slight benefits at the case level. Even conveying date of sale information at the case level would create an improvement in the benefit situation if it were performed globally. As a result of these friction losses, this concept for supply chain management is being implemented only sluggishly, as can be seen from the market structure study described in Chapter 3. In our view, in the near future it will be increasingly important for potential benefits to also be realized specifically in the upstream value processes of production and packaging logistics.

[1] See Section 5.4.

RFID for the Optimization of Business Processes   Wolf-Ruediger Hansen and Frank Gillert
© 2008 John Wiley & Sons, Ltd

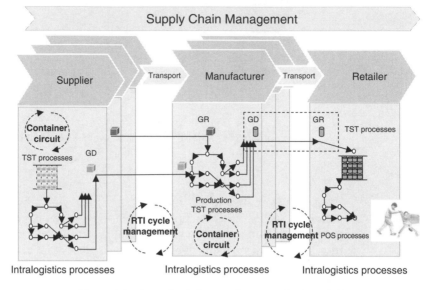

Figure 4.1   A supply chain spanning enterprise boundaries

In this chapter, we analyse the following business processes in detail and describe them with regard to the use of RFID technology:

- business processes in the retail and consumer goods industry;
- business processes in the packing industry;
- business processes in container cycles and returnable transport item cycles.

Here we show that the first two business processes are mutually complementary, with synergies that can lead to accelerated implementation of applications in the retail sector. The business processes in the third category also exhibit synergies and form a mutually complementary total concept.

4.1   Evolution from Supply Chain to Supply Net

Supply chain management involves integration of value chains that have global dimensions corresponding to trade routes. In parallel with this, it is necessary to establish a cooperative exchange of information between the enterprises involved in the value chain. Strictly speaking, supply chain management encompasses all value processes of the general order-to-payment process. This process includes all physical product production processes, order handling processes, product planning and control (PPC) processes with their derivative material demands and associated purchasing activities, and accounting processes with their associated monetary payment flows.

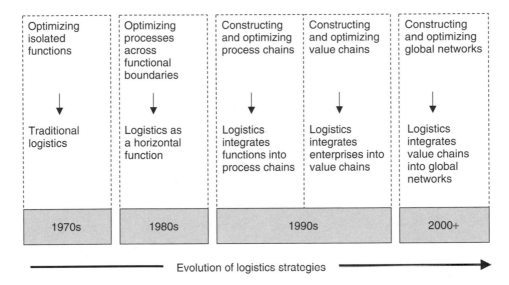

Figure 4.2 Evolution of logistics strategies according to Baumgarten/Walter (source: Bundesvereinigung Logistik eV; website http://www.bvl.de/70 1)

Figure 4.2 depicts the evolutionary development of logistics strategies, starting with the optimization of functional logistics subprocesses in the 1970s and extending to current global network strategies.

Progressive internationalization, changes in customer behaviour and shorter product life cycles are presently making sales markets even more dynamic. This increases the pressure to achieve global integration of the value chain. The supply chain is thus evolving into a supply net. New concepts for supply chain management have emerged as a logical consequence of this evolution, such as:

- JIT/JIS – just in time/just in sequence.

- VMI – vendor-managed inventory.

- ECR – efficient consumer response.

- CPFR – collaborative planning, forecasting and replenishment.

The current approaches alter the realms of responsibility in the value chain, and nearly all of these concepts are based on implementing comprehensive, innovative informatics support and centralizing strategic decisions concerning the entire supply chain. As both of these factors give rise to challenges that are not exactly trivial, it has not yet been possible to complete the process of integration across enterprise boundaries that was initiated in the 1990s. At the same time, the current situation in sales markets augments the bullwhip effect in the chain (disproportional amplification of stock levels in the upstream stages of the value chain due

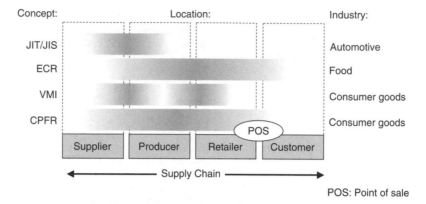

Figure 4.3 Four essential supply chain management concepts and their application areas (authors' depiction based on [Dan2003])

to small variations in demand). This phenomenon is described in more detail in Section 4.1.1.

Although the four concepts mentioned above help reduce the bullwhip effect, they also suffer from a lack of suitable IT infrastructures. As already indicated briefly in Chapter 2, RFID can give the discussion the impetus necessary to encourage the participants to make investments and institute changes.

Figure 4.3 shows the links in the supply chain that are addressed by the various concepts, or, in other words, the amount of integration depth achieved in the individual sectors. The integration width, which means which operational and/or strategic aspects are included, is not shown in the figure. It is described in more detail in the following sections.

4.1.1   Capabilities and Limitations of Supply Chain Management (the Bullwhip Effect)

The bullwhip effect (also called the Forrester effect) can be used to illustrate the mechanisms acting in a supply chain. This phenomenon has been known for 50 years already, and its causes and effects have been described extensively (see Section 2.1.1). The name comes from the shape of demand planning curves in the upstream links of a supply chain, starting with demand planning in a retail outlet. The curves look like a swinging whip. The amplitude of the swings becomes larger with each upstream link (see Figure 4.4). The mathematical basis for this phenomenon has been described by Helbing [Helbing2003] and Warburton [Warb2004], among others.

Here we can use an analogy from traffic research to illustrate the effect. Everyone is familiar with what happens on a motorway when traffic density keeps on increasing. Stop-and-go or standstill situations arise spontaneously, even when there are no obstacles to the flow of

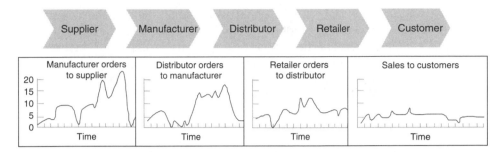

Figure 4.4   The bullwhip effect: amplification of order quantities by successive application of safety margins

traffic on the affected stretch. This arises from the fact that each driver responds to changes in the traffic situation according to his or her personal perspective. These individual responses are primarily influenced by the following parameters:

• The individual sensory capabilities and attentiveness of each driver, who personally determines the time delay of his or her response to visible events (e.g. the brake lights of the vehicle in front go red).

• Interpretation of the behaviour of the driver in the vehicle ahead – has that driver only braked as a precaution, or is a hazardous situation present?

• The estimated (forecast) evolution of the driver's situation based on his or her own experience – is the driver a commuter who is familiar with hazardous situations from his or her own experience, or does the driver respond with excess caution?

• Tactical considerations with regard to minimizing individual impacts – e.g. changing lanes.

All these parameters apply to the vehicle/driver subsystem, but they do not take the overall system (traffic flow on the motorway) into account. Drivers attempt to act according to their current information and their own experience in a way that minimizes personal risk. This leads to individual perceptions of safety margins for braking, and thus to traffic congestion without any visible cause.

Ultimately, what is missing is real-time information about the individual subsystems of the overall system and uniform rules for cooperation. Specific information from the head of the traffic queue would significantly improve the forecasting system in all vehicles further back in the queue. If each driver had access to the sensory data of the driver in the vehicle in front, braking behaviour would be optimized. In contrast to the present situation, this would reduce the severity of traffic congestion.

If all information were available in real time, only one significant factor would remain: the egoistic behaviour of individual drivers. If drivers continue to make rapid, unnecessary lane changes, the risk of traffic congestion and hazards will remain. In order to achieve an improved situation for all parties concerned, it is thus necessary for drivers to have a common desire to cooperate and subordinate their own interests to the collective interest of the group. This can only succeed if there is sufficient trust in the behaviour of other drivers.

If we apply this analogy to the bullwhip effect in the supply chain, we can see the following parameters:

- **Demand forecasts.** The supply chain participants forecast their future sales from previous key figures and add a safety margin to compensate for lead time. The retailer can keep this margin relatively small due to its proximity to the consumer, but this behaviour generates a cumulative effect in upstream feedback stages that produces large variations.

- **Shortage gaming.** If a shortage situation arises due to high market demand, the upstream stages will ration their products such that their customers, such as the retailer, receive products in proportion to their previous order quantities. The retailer will thus attempt to compensate for the feared reduction by increasing its current order quantity. This can cause the supplier at the head of the chain to drastically misjudge the market situation due to lack of information.

- **Order batching.** Discount incentives and reduced processing costs encourage customers to batch their orders. This batching amplifies the bullwhip effect.

- **Price fluctuation.** If a product is subject to wide price fluctuations, customers are tempted to build up reserves when prices are low.

Figure 4.4 shows the effect of successive amplification of order quantities in the upstream direction of the distribution chain. The relevance of the bullwhip effect (which has been studied for a long time) has increased because the amplitude of the variation at the producer end can now be derived from the variation in customer demand by using a mathematical model. This arises from changes in the market as described in Chapter 2. For example, shortcomings in sales planning become evident if we focus on the factor of changing customer behaviour.

Customer behaviour is characterized by three fundamental strategies:

- hybrid or multioptional behaviour;
- variety seeking;
- smart shopping.

Hybrid or multioptional behaviour refers to individual variance in purchasing behaviour. For instance, consumers may purchase their daily necessities at a discounter but treat themselves to luxury cars with expensive extras. By contrast, variety seeking behaviour results exclusively from a need for change. Smart shoppers act quasiprofessionally by concentrating on the entire economic context of pricing in the same way as a supply chain manager. Quality and market awareness are certainly predominant factors with smart shoppers, which distinguishes them from bargain hunters who focus on low cost. Smart shoppers utilize all available resources such as the Internet, bypass traditional links in the value chain such as specialist retailers, and buy directly from producers. The first two strategies listed above are demand-driven behaviour patterns. By contrast, smart shoppers are trained as well as served by the supply side. This quasiprofessional form of customer behaviour is roughly comparable to the methods of B2B processes [Werle2005].

Strategic cooperation in the supply chain is not enough by itself to reduce the bullwhip effect. With the currently perceptible nature of the effect, it is necessary for processes to adapt to the agility of the customers, and this adaptation is only possible with near-real-time information. Consequently, the business process optimization models described in more detail below can only be generally effective if they are based on near-real-time information, such as is possible with RFID.

4.1.2   Avoiding Traffic Congestion

Let us return briefly to the analogy of motorway traffic congestion and summarize our conclusions there. We mentioned two categories of parameters:

- individual behaviour of persons with decentralized strategies;
- missing or unrealistic information.

The behaviour of drivers on a motorway, which results from individual and situational circumstances, can be modified by two measures. The first measure is enlightenment, which means uniform education and training of the participants. In addition, rules must be specified and obeyed. An interesting example in this regard is the behaviour of drivers (and, in particular, German drivers) to a lane-end situation. The zipper principle, which stipulates that drivers should change lanes just before the lanes merge, evidently has the effect of minimizing delays. Nevertheless, suboptimal behaviour appears to predominate, perhaps due to a feeling of unfair advantage. The principle works well in conurbation areas where lane reductions are part of the traffic structure. By contrast, the previously described misbehaviour occurs at construction sites on motorways.

This reveals two aspects. First, commuters have a uniform view of the system from their daily experience and, second, supplementary signage usually helps drivers understand the situation better. Additional improvement can be achieved if the drivers know how much their behaviour and the behaviour of other drivers interact at any given moment. A 'simultaneous view' through the windscreens of other vehicles and the ability quickly to process perceived events quickly lead to overall optimization of the process.

If we transfer this analogy to the business processes of supply chains, we see that using RFID and IT systems across enterprise boundaries yields a view through the figurative windscreens of the other parties, which represents an enormous increase in trust for all concerned.

If we recall the ideas of ubiquitous computing and sensor networks and keep our analogy in mind, we can start to imagine what might be possible if RFID technology were supplemented by sensors or actuators. Driverless vehicles have long since left the realm of wishful thinking. Currently available top-end vehicles are already equipped with electronic distance control systems that produce continually optimized traffic flow. The combination of using autonomous control of logistics systems[2] for material flow subsystems for the purpose of tactical optimization and real-time information about the overall system for strategic optimization makes safe, reliable actions possible at high process rates.[3] What works for vehicle traffic works just as well for supply chain processes.

4.2   Strategies for Supply Chain Integration

In this section we describe individual strategies for supply chain integration and examine how they can be stimulated by using RFID.

4.2.1   Just in Time and Just in Sequence

The automotive industry was one of the first to recognize the advantages of supply chain optimization. Methods such as **just in time** (JIT) for logistics and production and **just in sequence** (JIS) have been in place for a long time already. The objective is to maintain minimal or almost no stock of incoming products in final assembly operations. Assembly stations draw the parts they need according to the pull principle, and the parts are only delivered to the assembly line at the time when they are installed. The prerequisites for effective operation of this principle are a steady demand curve and low model diversity.

The risk of production outages due to missing or improperly supplied parts, which is inherent to this principle, is minimized by zero-defect quality requirements and continual improvement processes (CIP). Long-term supply contracts and siting of individual infeed suppliers at the same location as the automobile manufacturer forge a close link via

---

[2] See Chapter 5.
[3] Traffic flow optimization is already possible with these telematics systems, which are not described here.

data synchronization using EDI. Nevertheless, there are still residual risks due to insufficient visibility of the physical material flow, especially with regard to increased model diversity resulting from increased mass individualization[4] (mass customization). For this reason, the JIT principle has evolved into the JIS principle, which means that parts are supplied to the assembly station not only at the right time, but also in the right sequence. An example would be supplying shock absorbers painted the same colour as the chassis in the correct installation sequence.

RFID can contribute to further quality improvement by ensuring that parts that cannot easily be distinguished from each other, such as dashboards,[5] can be assigned unambiguously to the right vehicles. Due to the high level of refinement of the JIT and JIS principles over the years, there is very little room for further improvement without using RFID.

4.2.2 Vendor-Managed Inventory

Vendor-managed inventory (VMI) is also a single-stage concept, and it is primarily used in the consumer goods sector.[6] Like JIT and JIS, VMI can be used in situations with small inventory variations in order to reduce the combined inventory levels of the two parties by having the supplier (the vendor) manage a physical or virtual inventory at the customer's premises. For this purpose, the supplier obtains a precise overview, usually daily, of the customer's inventory levels via EDI. Aside from the fact that this requires cooperation between the two parties, it makes the supplier responsible for the customer's inventory.

This assumption of responsibility is accompanied by the risk of reduced quality of the retrievable data. RFID can provide the necessary transparency here and mitigate the risk. The structure barrier is minimal due to the single-stage nature of the concept. The existing data link between the two parties provides a very good basis for implementation of RFID. Nevertheless, the benefits are limited by the shortness of the process chain due to the single-stage nature of the concept.

4.2.3 Efficient Consumer Response

Efficient consumer response (ECR)[7] has distinctly higher integration depth and width than the two previously described methods. The integration depth arises from incorporating all stages of the supply chain, while

---

[4] This refers to customized production of vehicles after receipt of order. Mass individualization is becoming increasingly available in the textile industry (mass production instead of alteration) and PC distribution (Dell).

[5] Dashboards with associated instrument panels can be configured with a wide variety of options, which can only be configured in the software state.

[6] VMI is also being used increasingly by tier-2 and tier-3 suppliers in the automotive industry.

[7] See GS1 (www.gs1.org).

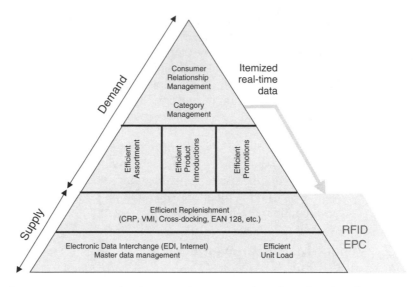

Figure 4.5 ECR pyramid and RFID application area (figure by authors, based on http://www.gs1.ch)

the integration width results from the extensive logistics and marketing modules. Figure 4.5 shows the structure of the ECR concept.

The upper part of Figure 4.5 relates to the marketing portion of the ECR concept. **Category management** comprises three vertical methods: product optimization, assortment optimization and promotion optimization. The top of the pyramid is the customer interface in the form of **customer relationship management** (CRM), which involves customer loyalty and customer care methods.

Data is extracted from the CRM system in order to supply information to the ECR tools. The three marketing methods are briefly outlined below:

- **Efficient product introduction.** A coordinated process for developing and launching new products in response to insufficiently satisfied customer needs. The objective of suppliers and merchants is to satisfy consumer needs at low cost.

- **Efficient promotion.** Coordinated organization and planning of sales promotion activities in order to minimize costs and increase effectiveness.

- **Efficient store assortment.** Having the right assortment is a decisive factor for key figures in retail trade. It correlates directly with inventory optimization and shelf space optimization, and it improves sales volume and sales per unit area. The sales volumes of producer and retailers can be boosted by having the right assortment. For this reason, out-of-stock situations are a predominant concern in retail trade.

This completes the description of the demand side of ECR. The lower part of the pyramid involves processes and standards on the supply side: CRP,[8] VMI and cross-docking[9] (processes) and EAN 128, efficient unit load,[10] EDI and master data (standards). It also includes the fourth core tool:

- **Efficient replenishment.** This integrates replenishment and follow-up deliveries between consumers, branch outlets, central distribution centres and producers. It is based on electronic data interchange such as described for the VMI concept. Functional reliability of efficient replenishment can only be ensured on the basis of accurate time and content information.

In order to ensure the availability of accurate time and content information, the objective is to achieve item-level data acquisition at the POS, which is made possible by RFID.

4.2.4 Collaborative Planning, Forecasting and Replenishment

Collaborative planning, forecasting and replenishment (CPFR) is based on a method developed by the Voluntary Interindustry Commerce Standards Association (VICS). It is intended to further integrate the previously described methods (ECR and JIT) and link them to the **enterprise resource planning** (ERP) concept. It relies on precise operational instructions.

The process of developing strategies for managing the supply chain exhibits continual change – sometimes in subprocesses and sometimes in further integration stages. In this regard, the objectives that can be derived from the bullwhip effect have not changed. CPFR is no exception to this, but it does clearly put the focus on effective collaboration, which is the weak point of all strategies. If efforts to improve the supply chain, enhance data quality and increase data quantity by increasing mutual trust among the participants are successful, full integration can also be realized successfully.

If the first factor is assumed to be given, the only remaining issue to be resolved is the quality of the information in the supply chain. This is why all hopes are based on RFID. The nine steps of the CPFR concept are shown in Figure 4.6.

---

[8] Continuous replenishment (CRP) can be defined as a high-level ECR strategy with the two variants: VMI and CMI (co-managed inventory).

[9] Cross-docking systems are pure transfer points in the supply chain, and they are described in detail in the description of business processes in the retail and consumer goods industries.

[10] Efficient unit load means modularization of load units and transport item in order to optimize the transportation, transfer and storage (TTS) process in the supply chain.

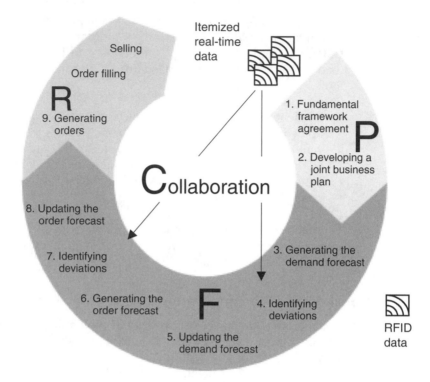

Figure 4.6   The nine stages of the CPFR concept

4.3   Business Processes in the Retail and Consumer Goods Industry

The trade component of the consumer goods supply chain fulfils two functions: bridging time by maintaining stocks in the distribution chain and bridging space by bringing goods close to consumers. Collecting and returning assets are also becoming increasingly important. In addition, the trade component provides the function of assortment formation by selecting and presenting goods according to demand from among the goods offered by producers, as well as the function of disaggregation, which means breaking down bulk quantities into consumer quantities suitable for household use. All of these functions are supported by marketing, advertising, promotion, consulting and customer service.

For the purpose of analysing the business processes, cash-and-carry markets occupy a predominant position in the fixed retail trade segment (self-service bulk stores, speciality shops, department stores, supermarkets, superstores and discounters) and the wholesale trade segment. Packaging has been the centre of attention of the Auto-ID Centers (Metro and Wal-Mart) since the start of the RFID discussion.

The essential challenges arise in the area of physical material flow due to the space bridging and disaggregation functions. Depending on the form of merchandising, stocks ranging from a few hundred items (discounters) to several tens of thousands of items (superstores, self-service bulk stores and cash and carry) are maintained, and they must all be purchased, sorted, handled and presented at the right time. The time bridging function is minimized by replenishment. Knowledge of where individual items are available in which aggregation states is crucial for physical provision of the goods to consumers. It is generally expected that RFID will primarily help avoid **out-of-stock** situations, which means gaps on the shelves in retail outlets. According to ECR Europe, the impact of these situations on brand-name manufacturers is that nearly 40% of consumers choose a different brand, and the impact on retailers is that more than 20% buy from a competitor.

In 2004, the **Global Commerce Initiative** (GCI)[11] issued an 'EPC roadmap' describing the potentials of the RFID supply chain (see Figure 4.7). However, full utilization of all of these general potentials requires RFID penetration at the item level. It is difficult to judge the extent to which these potentials can already be utilized at this level. In any case, the potentials assigned to the factory can only be realized at the item level. In the producer's finished goods inventory, benefits can arise already at the case level due to reduced labour costs. There is a clearly increased benefit in trade distribution (distribution centres) due to faster processing of goods receiving and shipping. In the functional area of branch logistics, labour cost reductions can be expected at the load unit level. Larger potentials can only be found at the item level. This includes reduced customer pilferage. A positive effect on pilferage in the

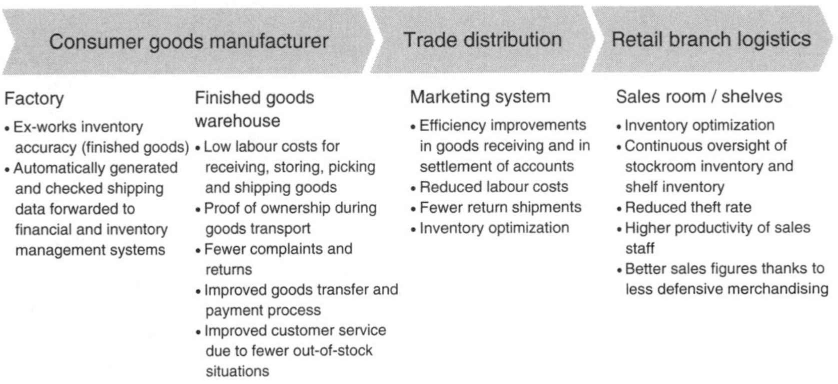

Consumer goods manufacturer		Trade distribution	Retail branch logistics
Factory	Finished goods warehouse	Marketing system	Sales room / shelves
• Ex-works inventory accuracy (finished goods)	• Low labour costs for receiving, storing, picking and shipping goods	• Efficiency improvements in goods receiving and in settlement of accounts	• Inventory optimization
• Automatically generated and checked shipping data forwarded to financial and inventory management systems	• Proof of ownership during goods transport	• Reduced labour costs	• Continuous oversight of stockroom inventory and shelf inventory
	• Fewer complaints and returns	• Fewer return shipments	• Reduced theft rate
	• Improved goods transfer and payment process	• Inventory optimization	• Higher productivity of sales staff
	• Improved customer service due to fewer out-of-stock situations		• Better sales figures thanks to less defensive merchandising

Figure 4.7   General RFID potentials (figure by authors, based on [GCI2005])

[11] The Global Commerce Initiative is a voluntary body that was founded in October 1999 with the aim of enhancing the performance of international logistics processes for consumer goods by joint development and recognition of recommended standards and improvements to core business processes. Its member organizations include GS1 and VICS.

supply chain can already be seen at the pallet level, and it is even more pronounced at the case level:

Shrink Reduction in the Supply Chain
Industry estimates shrink levels to be approximately 2 percent of sales worldwide. Today retailers are tagging pallets, dollies, cases, and trays to track these units internally within their own supply chains. IBM business case analysis shows that the use of RFID can improve shrink for the average retailer by 25 percent at the case level and up to 40 percent at the item level.
GCI EPC Roadmap

As shown in Figure 4.7, in addition to process cost optimization there is a particular potential for reducing transaction costs at all stages by automatic generation of shipping data, proof of ownership during goods transport, improved payment transactions, and so on.

In order to analyse the potential benefits more exactly, it is necessary to break down the processes in more detail. Figure 4.8 depicts the consumer goods supply chain process from producers to consumers. It shows the essential subprocesses as well as the main processes of consumer goods manufacturing, trade distribution and branch logistics. The main process in the packaging industry is described in Section 4.4, and the transportation process for returnable packaging is described in Section 4.5. An exact analysis of the retrologistics[12] processes of collecting and returning assets and equipment would exceed the scope of this book, but this does not mean that the subject is unimportant.

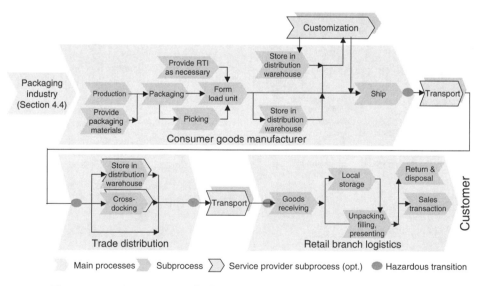

Figure 4.8   Main processes and subprocesses of the consumer goods supply chain

---

[12] A modern term for 'disposal logistics'.

The processes necessary for packing and forming unit loads depend on the goods produced by the manufacturer. Unit loads are the handling units for each subsequent process operation, regardless of whether the process involves direct shipment or storage. Custom packaging (special handling of items) may be part of the process chain. This is often done by service providers (contract packaging) to produce special versions or special packaging, which is fed back into the manufacturer's process chain. The distribution warehouse is normally the source and destination for custom packaging.

In many cases, products are picked to form single-branch pallets that are shipped as full loads. In the shipping department, each unit is labelled with a barcode (EAN 128) and a Serial Shipping Container Code (SSCC) in accordance with the standardized terminology of GS1. From this point on, full loads are transported via an optimized supply chain to a cross-docking point if direct delivery is not a reasonable option. The carrier may be the manufacturer, the retailer, or in many cases a service provider. Cross-docking points are part of the ECR concept, and they support regrouping single-branch full pallets from manufacturer-specific full loads (A, B and C) to form single-branch full loads (a, b and c). Figure 4.9 shows the operating principle of cross-docking.

Cross-docking points are often located in distribution centres with storage, but no storage is associated with cross-docking. This is exactly the optimal case. Furthermore, no picking is involved in classic cross-docking, so the cross-docking point can be assumed to perform a pure transfer function.

If this is augmented by a picking function, the term 'transshipment point' (TSP) is used (see [Gab2004]). In this case, homogeneous full pallets from the manufacturer are broken open and items are picked

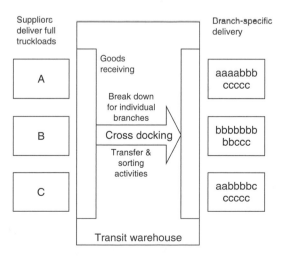

Figure 4.9   Cross-docking [Gab2004]

to form full or partial single-branch pallets. Both of these processes are described in more detail in Section 4.3.1.

The next transportation process bridges the distances to widely distributed consumers. In the German retail sector, goods are supplied to 264 retail outlets for every 1 million inhabitants in the food sector alone (for comparison, in France the number is 117.3). The total number of retail outlets in the food sector with floor areas greater that 400 m^2 operated by German retailers is thus approximately 30 000. Combined with approximately 20 000 different items and regional differences in assortments, this quickly leads to a high degree of complexity.

4.3.1 Cross-docking and Transshipment

Cross-docking and transshipment are key process elements in the supply chain, and they warrant being described here in more detail. Besides the physical material flow tasks (the transportation and transfer functions), this can indicate the basic informatic conditions necessary for RFID use and potential RFID applications. The basic principle of cross-docking and the extent of value aggregation of objects in the material flow, which has a decisive impact on process costs, are shown in Figure 4.10.

The distribution chain can be regarded as an aggregation and disaggregation stage for material flow objects. Transportation, storage and transfer (TST) processes are performed at the highest possible aggregation levels defined by standardized load and transport units. Full pallets with standardized dimensions (e.g. 1200 × 800 × 1950 mm for CCG2) and containers for full loads are essential parts of the distribution chain. For cross-docking, the manufacturer supplies prepicked, single-branch

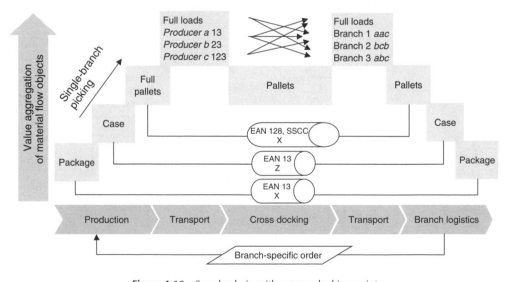

Figure 4.10 Supply chain with a cross-docking point

full pallets. Picking can be performed during item aggregation without any need to first break open previously formed units and resort them. A fundamental prerequisite for this process is timely, error-free transmission of branch-specific order data from retailers so that picking can be included in the overall process during work preparation. Implementation of distribution at the highest possible aggregation level using uniform information (source/destination identification of item, case and load unit information) along the entire chain presupposes optimized process execution by the manufacturer so that process and transaction costs can be minimized. Retailers must also fulfil stringent conditions if this concept is to be used. The order quantities for each branch must be large enough to make pallets and full loads reasonable. The cross-docking strategy is not suitable for relatively small branches, follow-up deliveries and the like. Consequently, cross-docking points are often combined with storage operations and other distribution strategies.

The RFID potentials that can be exploited in this connection lie in accelerating receiving and shipping at the pallet level at cross-docking points and improving the operation of these processes. The importance of this aspect should not be underestimated, since every time-bridging function necessary at a cross-docking point unavoidably leads to storage, which reduces efficiency at this point and should therefore be avoided. Keeping individual load units intact over the entire supply chain allows data flows to be simplified, which in turn improves their quality. For instance, goods receiving at the branch loading dock can be supported by automatic functions and compared with the despatch advice messages (DESADV) from the producers.

Distribution within the selling area, which means branch logistics, can only be supported by RFID if at least the cases (and preferably individual items) are fitted with RFID tags. At the producer end, there is a similar situation because the producer's highly automated, informatically closed aggregation processes for fast-moving consumer goods (FMCG) exhibit fewer transparency gaps.

As already mentioned, a cross-docking strategy can only be effective if it is not necessary to maintain any stocks at the cross-docking point. As it is not always possible for the producer to perform final picking at the full-load level, the cross-docking concept can be expanded to the transshipment concept by adding a picking step to the pure transfer function. As with cross-docking, replenishment is also avoided here. Transshipment is portrayed in Figure 4.11 in the same manner as for cross-docking.

Here the producer assembles load units containing only one type of item, which means that load units are formed based on production instead of orders.[13] These homogeneous full pallets are labelled with EAN

---

[13] In the first approximation it is not order-specific. Naturally, in connection with ECR, CPFR and so on, all production activities should be based on sales information.

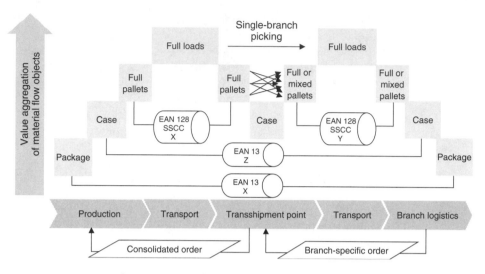

Figure 4.11 Supply chain with a transshipment point

128 barcodes and SSCCs and shipped with despatch advice messages as previously described. Cases and items are also EAN labelled as described in the previous example. In contrast to classic cross-docking, this process falls short of optimum aggregation levels in two regards: full pallets must be broken open and partial load units are formed afterwards.

Here again, RFID can have a positive impact on the time aspects of goods receiving and shipping at the transshipment point if the pallets are equipped with supplier tags and SSCCs (EAN 128 barcodes). Further improvement of the processes at the transshipment point can be achieved if RFID is used at the case level.

After the full pallets are broken open, cases must be resorted in the picking process and new load units must be formed. The first significant transparency gaps occur in this process, and they lead to loss or theft of cases. Automated identification at various points in the process would help close these gaps and thus make inventory management unnecessary.

The transaction costs of the order process are higher than with cross-docking because a two-stage process is necessary. Orders for individual branches are consolidated at the transshipment point and conveyed to the producer as combined orders. This intermediate step harbours the risk of demand amplification as described in Section 4.1.1. This can be avoided by using a closed information loop in a CPFR system. Here it must be borne in mind that cross-docking and transshipment points need not necessarily be operated by the merchant; they can also be operated by service providers offering various ranges of services.

4.3.2   The Role of the Logistics Service Provider

Current logistics service provider concepts differ in four regards:

1. Are the provided services operational or administrative?

2. Does the service provider use internal or external resources?

3. Are complete logistics services provided to the enterprise?

4. Is the supply chain integrated internally or end-to-end?

These four aspects involve the following considerations:

- **Operational versus administrative services.** Carriers typically provide operational services, while freight forwarders also provide administrative services.

- **Internal versus external resources.** The service provided by a carrier is based on providing transportation using the carrier's own vehicle pool. A carrier can also work together with a freight forwarder, which need not necessarily use its own resources.

- **Complete logistics management.** This is undertaken by a contract logistics services provider, which is also called a third-party logistics provider (3PL). It can operate with or without its own resources.

- **Internal versus end-to-end supply chain integration.** This is what is called a 'fourth-party logistics provider' (4PL). The role of such a provider is to control and coordinate the entire supply chain. This also includes selecting and operating the information systems. A 4PL provider, which is often a joint venture of the parties in the supply chain, thus serves to institutionalize the ECR and CPFR strategies.

With regard to RFID, logistics providers have different roles depending on their orientation (see Figure 4.12). Places where handling operations such as picking, packing and labelling are necessary are especially

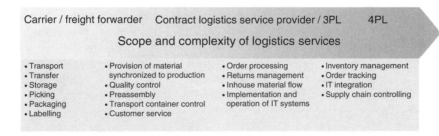

Figure 4.12   Core tasks of logistics service providers

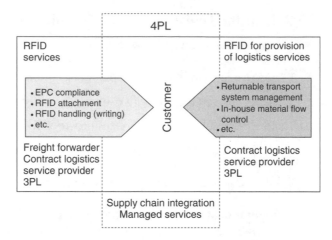

Figure 4.13   RFID as a service and as an optimization strategy for logistics service providers

suitable for ensuring EPC compliance,[14] especially during the start-up phase of implementing requirements imposed by retailers. As part of container management tasks, service providers can use RFID to support their internal processes (see Section 4.5). This makes RFID important for logistics service providers (such as the packaging industry as described in Section 4.4) as a dual source of added value.

Figure 4.13 shows the relationship between RFID services and using RFID to optimize core logistics services.

For a logistics service provider, providing RFID services can serve as an entry point for migration to extended uses. In particular, 4PL providers will have an opportunity in the future to expand their portfolios to include the data volumes accumulated as part of supply chain integration as a managed service (see Section 6.3.2).

4.4   Business Processes in the Packaging Industry

The 'internet of things' concept has two components: an informational component (the internet in the virtual world) and a physical component (objects in the real world). RFID connects the two levels by bridging the media discontinuity. The architectures in the virtual world are already quite advanced (see Chapter 5). For the economically necessary penetration of RFID down to the item level on a broad front, more attention must be given to using RFID on or in objects. Up to now, the discussion has been conducted two levels: standardization (of RFID protocols and virtual identities) and integration (of IT structures). The operational challenges facing

---

[14] EPC compliance means fulfilling the requirements of the retail enterprise as defined in its mandated policies. For example, this includes 'slap&ship' in the context of pallet identification.

the realization of the internet of things are considerable. Consequently, the international Auto-ID Labs and producer consortium EPCglobal are working intensively on developing the necessary structures and standards.

The issues here are reminiscent of **electronic article surveillance** (EAS) systems, which were proposed in the early 1990s as a form of source tagging for integration at the production stage. However, introduction of such systems was a drawn-out process and occurred only after many years of international debate among the enterprises concerned.

4.4.1 Lessons from the Source Tagging Concept

Introduction of source tagging was promoted especially strongly by large retail chains such as Wal-Mart and Home Depot in the USA. The objective was to reduce customer theft, which still accounts for half of all inventory discrepancies. Source tagging was seen as a way to improve the cost situation by automated tagging during production instead of manual tagging in retail outlets and achieve better results due to the fact that the tags were hidden. This requirement confronted producers of consumer goods with many new aspects that had an impact on their business processes. As with the present situation, in which it is not necessary to tag every item or all items of a group because some customers do not impose this requirement, it was necessary to introduce logistics measures to handle two versions of each item (two item numbers, two storage locations, etc.).

The **Consumer Product Manufacturers Association** (CPMA)[15] was founded in 1999 with the objective of making retailers aware of the impact on the business processes of the manufacturers. Standardization became a focus of consideration, and it strongly shaped the orientation of the Auto-ID Center at the Massachusetts Institute of Technology (MIT) in Boston, USA.

In addition to resolving logistical and organizational issues, the technological obstacles to integration of EAS into products and product packaging also became a subject of concern. The first German studies on this subject were initiated in 1993 by the Logistics Department of the University of Dortmund. Studies on source tagging and VDI guidelines formed the basis for further research activities in the area of intelligent packaging [Gil1994; Gil2001; VDI]. They were the first studies of the impact of tagging activities in the consumer goods industry and the upstream packaging industry, and they form the basis for further studies on RFID. On the one hand, integration of RFID into products or packaging has an impact on the business processes of the packing industry and manufacturers of packaging materials. On the other hand, this integration

---

[15] The CPMA was founded by Eastman Kodak, Johnson and Johnson, the Gillette Company and Procter & Gamble. It can be regarded as one of the precursors of the Auto-ID Center.

creates new business processes for attachment and handling operations (writing EPCs and administering number pools) that create added value for customers. This insight is already widespread in the packaging industry, and thus among producers of consumer goods, but it is not yet sufficiently recognized in the upstream value processes.

RFID implementation in the consumer goods value chain is driven by the mandated requirements of retailers such as the Metro Group. This starts with tagging at the pallet level, which provides benefits for the retailer but not for the suppliers. RFID tagging at the case level was planned to be introduced in the course of 2006, and it will also yield benefits for suppliers. Tagging at the item level is even more attractive because it will yield distinct benefits for suppliers if the information obtained from RFID reading activities extending as far as the POS is also made available to suppliers.

Large-scale use of RFID tags must be economically viable. Consequently, the tag price must drop to a suitable level, which is typically in the range of 1 to 5 cents. Studies on source tagging have shown that use of such components can have a major impact on existing production processes. A low tag price is thus not the only criterion; attachment costs and process adaptation costs must also be considered.

This can best be done by performing a total cost of ownership (TCO) analysis for the object to be identified. An object with an RFID tag *receives* RFID functionality and is *handled* according to how the tag is attached. The costs of the RFID tags and handling operations must be evaluated and compared with the overall benefits. This may show that the RFID project would be economically viable even with more expensive tags, but it may also show that the project is not economically viable even with less expensive tags because the process costs are too high.

4.4.2   The Route to Intelligent Packaging

The term 'intelligent packaging' provides a framework for the previously mentioned aspects of economic integration of RFID functionality into objects and indicates that it offers the packaging industry an opportunity for supplementary added value. Here we examine the business processes of the packaging industry in more detail as an example and describe suitable realization routes. Current applications in the consumer goods chain are almost exclusively located at the pallet level. Even here, labelling and coding pallets (which are also a form of packaging) as load units is a rudimentary form of intelligent packaging and the first step in a migration process. The load formation process is shown in Figure 4.14.

The packager (the producer of the consumer goods) puts together the loads, usually in homogeneous form – which means that only a certain number of items of a particular type are on the pallet. If an RFID tag is attached at the pallet level, this currently takes place during the load unit

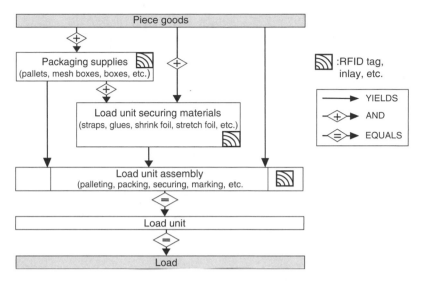

Figure 4.14   Formation of load units and loads with options for using RFID [Bode2004]

formation process. In particular, a tag that has been printed and coded in a printer is attached to the side of the pallet at a height of approximately 80 cm after the load unit has been secured with stretch film.[16] However, the RFID function could also be integrated in the load unit securing material (e.g. in a strap) or in packing aids such pallets.[17]

In the case of returnable transport item systems, which are described in detail in Section 4.5, integrated RFID elements have excellent economic potential because they can be reused, and they do not incur any attachment costs in use. Depending on the type of process used by the consumer goods producer, the case-level packaging process can also be regarded as load unit formation. However, it is also possible to regard case-level packaging as part of the piece goods formation process depicted in Figure 4.15. This depends on the level of integration of the process (a packaging area in the picking department versus a fully automated packaging line).

Applying tags at the item level leads to considerably increased complexity due to the following conditions:

- Different types of packaged goods (solid, liquid or gas) require different types of packaging materials and packing aids:
    - Integration of RFID functionality is possible only to a limited extent with tubes and bags due to size and stability issues.

---

[16] The Applied Tag Performance Working Group of EPCglobal focuses on recommendations for the point of attachment. Current results can be obtained from or consulted at GS1.

[17] See, for example, the Craemer intelligent pallet:
http://www.craemer.de/Craemer_Kunststoffpalette_CR1_mit_Transponder.htm.

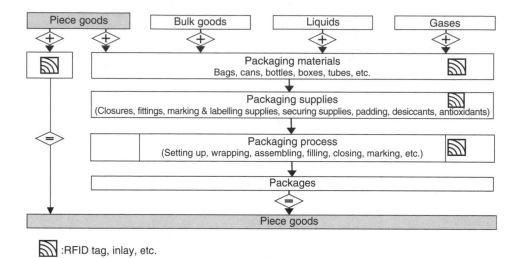

:RFID tag, inlay, etc.

Figure 4.15 Formation of piece goods with options for using RFID [Bode2004]

   - Operation of RFID devices can be impaired by packaged liquids
     and metal cans if suitable countermeasures are not taken.

• Different processes in piece goods formation require precise process
  analyses:

   - Speed of the packaging process.

   - Stability of the process (susceptibility to disturbance).

   - Specific packaging process costs.

Nowadays barcode labelling is usually done during package printing.
At present, RFID functionality must be achieved by using a substrate (inlay
or tag), and the substrate must also be written electronically (e.g. with
the EPC item number). The packaging process must always be analysed
precisely in order to determine the lowest TCO level. This depends on
the following aspects:

• Are the modifications to the packaging line economically viable, or
  would it be more reasonable to perform RFID attachment as separate
  operation?

   - How many items or groups will be equipped with RFID tags during
     the time frame of the analysis?

   - What percentage of the total volume of a particular item will be
     fitted with RFID tags?

• Are the potential impacts on the packaging process calculable (see
  Figure 4.16)?

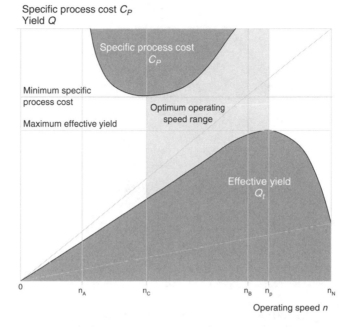

Figure 4.16 Process control in packaging [Bleisch2003]

– How will the effective yield of the packaging line (Q_t) or the working speed change in response to the process modification?

– How will the specific process costs (C_P) change?

– How will the optimization target figures change (reliability, quality and flexibility)?

The crucial changes to business processes in the packaging industry due to using RFID functionality can be expected to be found on the process cost side and in the quality area. Even the present 'slap & ship' concept at the pallet level should more properly be called the 'slap & encode & verify & lock & ship' concept, since it is necessary to verify that the right EPC has been written and check for correctness and proper operation.

All of this essentially leads to the following basic process requirements for quality assurance of RFID functionality:

• damage-free attachment of RFID components;

• functionality verification;

• coding (e.g. EPC) – correct code from correct number pool;

• verification;

• blocking write access.

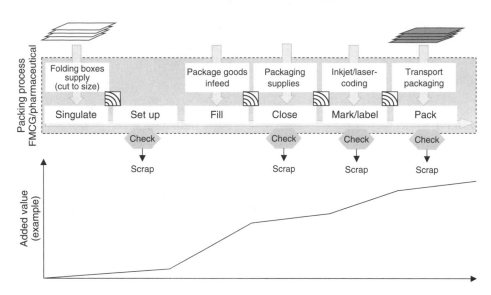

Figure 4.17   Value accumulation during the packaging process (qualitative representation)

These RFID attachment steps must be appropriate to the working speed and must occur very quickly, which is a decisive factor for optimum implementation of the process steps (singulation, assembly, etc.). Consumer goods manufacturers cannot achieve the intended objective here without the assistance of specialists in RFID and automation. Current RFID systems are designed for various forms of attachment to allow economically viable unit costs to be achieved despite low quantities. If RFID were an integral part of packaging lines, like PLCs, the next decision criterion would be the reject rate in the selected process step. Figure 4.17 shows an example of added value per process step. For example, a reject represents a larger value loss if the tag is coded and attached after container filling. Nevertheless, this process step may be a reasonable attachment point due to available working speed. Based on these insights, it is obviously necessary to consider attaching RFID tags as close as possible to the start of the value chain.

4.4.3   The Route to Intelligent Packages and Packaging Materials

The following descriptions are related to the subject of source tagging. Already at that time, analyses were generated to assess the feasibility of integration as close as possible to the packaging material production stage. The paper, glass and plastics industries were also included in the discussion in order to examine the possibility of upstream integration. Due to the roughness of the processes and the temperatures and pressures used, current RFID components are relatively unsuitable for integration into this process step. The absence of exact information about the intended

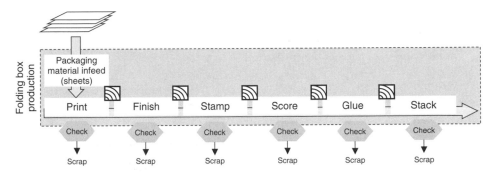

Figure 4.18 Qualitative value generation in the package production process

use of the basic materials also restricts the scope of use here. Each RFID component must ultimately be part of a consumer package, such as a perfume box, which can be made from the same lot of cardboard as any other type of folded box. Although we should mention that efforts are being made to find a solution to this problem, it would require very inexpensive and robust RFID components. In the medium term, the route to success is integration into the actual packaging, and in particular into folding boxes (Figure 4.18).

As with the packaging process, each subprocess of the package production process represents a possible attachment point with corresponding inspection requirements and rejects. Here again, the lost value represented by rejects increases along the process chain.

Commercial integration will make new RFID construction methods necessary. The present widely used process chain for discrete tags depicted in Figure 4.19 will give way to technologies that use the packaging material as a substrate. Current efforts are focused on printable electronics, with polymer-based methods being only one of the options. Directly printable antennas based on metallic components are also conceivable.

The rate of integration in the upstream links of the value chain depends on several factors:

- development of innovative chip bonding technologies;

- methods for printable antennas suitable for use on an industrial scale;

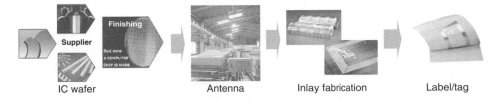

Figure 4.19 The value chain of a conventional tag

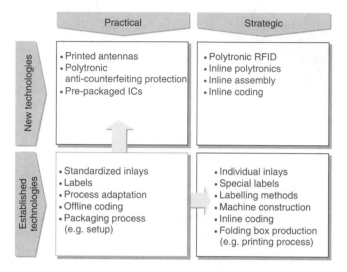

Figure 4.20 Solution matrix for intelligent packaging

- developments in polytronics;

- development of sound, shared business models for package manufacturers and the packaging industry;

- Evolution of the total volume of RFID applications.

Particularly with regard to introducing RFID into trade and industrial processes, there is no royal road to success. Like all other integration processes, it will be determined by specific starting scenarios and migration steps appropriate to individual tasks and requirements. Figure 4.20 shows some suggestions for arriving at solutions. Current technologies and methods should be examined with regard to their suitability for backward integration. To this end, all participants in the process chain should meet together and devise solutions. Just as with forward integration in the ECR and CPFR concepts, trust-creating measures are necessary because each participant in the process attempts to optimize its own part of the process. Packaging material manufacturers must be prepared to modify their processes, while the packaging industry must be prepared to share the costs.

A service-oriented packaging industry will regard it as an advantage to be able to provide additional services that fit well with existing activities. Backward integration of coding tasks (such as EAN 13 or best-before dates) and specific texts often forms part of a particular package of services, and expanding this package to include attaching RFID tags and performing the necessary coding, such as with EPC numbers from a suitable number pool, will be regarded as innovative. Optimization of business processes between packaging material manufacturers and the packaging industry is possible, but RFID-related customer services primarily represent a new business processes for the packaging industry.

4.5 Business Processes for Container Systems and Returnable Transport Item Systems

4.5.1 Cyclic Processes and their Control Requirements

Container systems and returnable transport item systems are more or less closed economic systems within supply chain processes. The simplest case occurs when they are used in the intralogistics of an enterprise. The structures are more complex when they involve a pool managed by a service provider.

Figure 4.21 shows the general structure of a pool system with a cycle starting with outgoing full containers, extending to the consumer via transport processes, passing through another transport process to the pool service provider, and finally returning to the producer. Particularly at the service provider node, cycle-specific value is added in the form of cleaning, repair, storage and sorting.

The term 'returnable transport item' (RTI) covers a wide variety of containers and packing aids, such as boxes, pallets, mesh boxes, racks, and so on. Sea cargo and air cargo containers can also be regarded as a form of RTI. The cost ranges from a few euros for a simple plastic small-load carrier (SLC) to several hundred or even many thousand euros for special racks for the automotive and aviation industries. Figure 4.22 shows several examples of RTI systems classified by type of use (open or closed cycle).

The necessary standardization activities are carried out by the relevant industry associations, such as the German Association of the Automotive Industry (VDA), or GS1.

The importance of returnable transport items (RTIs) increases as the cost of disposable packaging is increased by regulations for mandatory taking back and recycling. The essential advantages and disadvantages are shown in Figure 4.23, which illustrates the higher demands placed

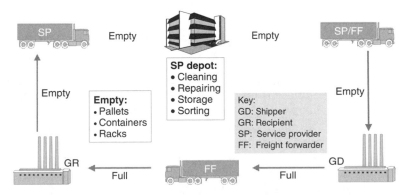

Figure 4.21 Economic cycle of a returnable transport container system

	Proprietary	Industry-specific	Open
Containers	• Karstadt return system • TNT Pallecon system • bi-Box modular system • Returnable containers	• Euro meat boxes • Returnable glass bottles • Vegetable and fruit crates • Returnable fish boxes	• Collico container system • Tengelmann returnable system (RTS) • VDA KLT containers • Logstar system
Pallets	• Slip sheets • Inka returnable processed-wood pallets • Inka returnable plastic pallets • remaplan pallet system		• Euro pool pallets • CCG concept • Chep half pallets and quarter pallets

Figure 4.22   Comparison of RTI systems (authors' depiction based on [Bode2004])

	Disposable	Returnable
Advantages	• Lower production costs • Lower transportation costs • No financial or energy expenditures for cleaning • Specific packaging for specific requirements • Can be fashioned individually	• Less environmental impact from production • Improved goods protection due to stronger construction • Pooling reduces transport effort for empties • Complies with legislative expectations • Rising consumer acceptance
Disadvantages	• Environmental impact from production and discarding • Disposal problems • Declining consumer acceptance • Take-back and recycling obligations	• Ties up considerable capital at present • Redistribution structures must be developed • Costly administration (including repairs) • Cleaning and sorting of returnable containers • Uniformity diminishes individual identity

Figure 4.23   Advantages and disadvantages of RTI relative to disposable packaging systems (authors' depiction based on [Bode2004])

on managing returnable packaging systems. Optimal management of the specific RTI business process is a key to its economic viability, and it requires suitable information technologies.

In many cases, the modular design and versatility of the containers leads to shrinkage due to pilferage and use for unintended purposes. For example, SLCs are often used to store files in offices or kept in the boots of cars as handy containers. The net effect of this cycle leakage, regardless of the reason, is supplementary purchases with associated additional costs. Lack of transparency also leads to excess stocks, which ultimately means higher capital commitment. Another thing that is often seen in pool-based systems is that users keep good containers for their intralogistics processes and put bad ones back in to the cycle immediately. This generates an increased load on control and management functions. In summary, the following specific tasks can be identified for RTI management:

- Stocktaking and control functions.

- Overview of circulating inventory and circulation rates.

- Traceability.

- Shipment tracking (tracking and tracing).

- User-pay cost accounting and billing.

- Integration of all process chain participants (kanban containers).

All aspects of the RTI cycle, including production, shipping and traversing the customer organization, must be considered in order to see it from a specific enterprise perspective (Figure 4.24). Here we propose three RTI management task areas that can be supported by RFID.

- **Inventory management:**

  - maintaining master data (container type, manufacturer, volume, etc.);

  - procurement (replenishment and expansion);

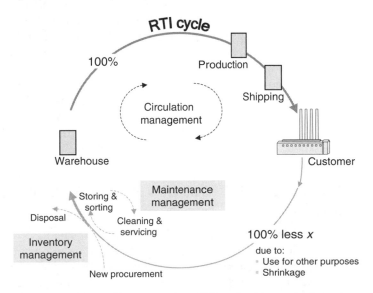

Figure 4.24   Returnable transport item (RTI) cycle without a pool structure

  – reassessment as part of asset valuation;

  – disposal (old containers and irreparably damaged containers);

  – storage, sorting and supply.

- **Maintenance management:**

  – container cleaning (possibly by a service provider);

  – repair (possibly by a service provider);

  – modifications (e.g. of car racks in the context of a model facelift);

  – life cycle documentation.

- **Circulation management:**

  – circulation figure and rate;

  – dwell time at each destination (e.g. customer or production);

  – availability of specific containers at each storage location.

Naturally, process optimization aspects related to using RFID should not be ignored. For instance, the possibility of acquiring data for an entire load unit of outgoing goods (bulk acquisition) harbours considerable potential benefits. Similar process benefits can also be obtained from RFID when the containers are returned. However, reliable bulk acquisition is only possible if the RFID tags are optimally positioned on the containers. Suitable studies and recommendations in this regard are available, such as **Applied Tag Performance** (ATP) from the EPCglobal group. The factors that affect the identification rate, such as metals, liquids and tag overlapping, are described in detail in Chapter 7. Here we only mention that there are RTI systems available in which the container volume can be reduced for transporting empty containers. This includes folding and nesting containers, for example. If the tags are positioned optimally for the full-container situation, this can lead to tag overlapping when empty containers are shoved together and thus to degraded identification during bulk reading of the empties.

4.5.2 Economic Viability Analysis of Using RFID with Returnable Transport Item

A sample cost/benefit analysis using the key figures of a specific RTI system is described below. It is based in part on a presentation by Strassner, who prepared it as part of a dissertation at the University of St Gallen dealing with the activities of M-Lab [Strass2005]. It is also based on the authors' infrastructure cost assumptions based on known applications.

The example shown in Table 4.1 is based on a life cycle analysis. The costs and potential benefits are referenced to a RTI service life of four years in this example. The estimated cost of RFID tagging is €0.71. In light of the requirement that the tag be washable, this is a realistic value. The number of cycles in the third row of the table is shown as increasing from 10 to 12. The two additional cycles per year for the container could

Table 4.1 Economic viability analysis of an RFID-assisted SLC system [Strass2005]

RTI Key Figures		Actual	Transponder-Assisted
1	SLC service life (year)	4	4
2	SLC cost (€)	5.00	5.71
3	Average number of cycles per year	10	12
4	Average number of SLCs per pallet	40	40
5	Average number of labelling operations per cycle	1–2	0
6	Time required to apply label (min)	0.2	0
7	Time required for incoming goods inspection of 40 SLCs (min)	5	0
8	Random check rate	1/50 pallets	Each time
9	Annual shrinkage rate (%)	5	3
10	Cost of 1 minute of labour (€)	0.35	0.35
Savings over 4 years (RTI service life)			
	Savings on incoming goods checking (€)		0.035
	Savings on labelling (€)		4.20
	Savings on shrinkage (€)		0.40
Additional cost for RFID use (€)			**0.71**
Net benefit over 4 years (SLC service life) (€)			**3.93**

also be expressed in monetary terms by a change in capital commitment due to reduced inventory. The corresponding analysis is not included here, but it does harbour an additional potential benefit. The fourth row relates to the process. The number of containers per pallet is shown as 40. As no distinction is made here between transporting full containers and empty containers, it can be assumed that the containers are stackable but not folding or nesting SLCs, since otherwise the number of RTIs per pallet would be correspondingly higher on the return trip. The fifth row relates to a frequently underestimated benefit of RFID technology in bridging IT infrastructure boundaries. Barcodes are often selected by each enterprise individually. Effective standardization such as that generated by GS1/EPCglobal for the consumer goods chain is absent in other industry sectors. Relabelling in the supply chain is thus common practice, and the value of two relabelling operations per cycle used here is on the conservative side.[18] The sixth row shows the average time needed for labelling, which is used to calculate the resulting cost. The figure '5 min' in row seven shows quite clearly the effect of process improvement by bulk reading in place of individual identification. However, this does not have a significant monetary impact because random checking is limited to every 50th pallet. Here another nonmonetary potential can be expected with regard to quality assurance. The 40% reduction in shrinkage can be regarded as a conservative estimate if we assume that use of RFID will give the user-pays principle teeth by providing evidence of where items disappear from the system.

The total benefit over the service life of the containers is €3.93. In this example, the potential savings arise primarily from avoiding relabelling. This can vary from one case to the next, and it depends on the actual business processes. This variation is illustrated by the following results from another pilot project.

Pilot project in a production plant, based on 10 000 containers[19]

- Circulation rate increased by up to 5–8%. A 20% increase in the rate was achieved in the previous example. This could mean that either circulation management was better organized in this pilot project or the overall complexity (such as purely internal logistics) was lower.

- Shrinkage reduced by up to 3%. This value is comparable to the value in the previously described example.

- Effort for searching for containers reduced by up to 75%. From this it can be concluded that the impact of a missing container is very large. See also the last point (reduced production idle time).

---

[18] Naturally, the potential of RFID appears in a different light if we consider stable barcode labels that survive several use cycles.

[19] Source: Silverstroke AG.

- Misrouted items reduced by up to 95%. See the comment on the previous point.

- Production idle time due to missing containers reduced by up to 35%. Here the potential benefit does not arise from handling large numbers of containers, but instead from avoiding relatively rare outages with major impacts. This might be an example from the automotive industry, where zero defect level has high priority.

In summary, it can be said here that potential savings can only be revealed by a thorough process analysis. Prediction of specific savings potentials in the RTI area is only possible to a very limited extent. However, there are certainly individual aspects that are always effective, such as reducing stocks and increasing circulation rates.

Now that we have examined the potentials per container, let us fill out the picture by looking at the infrastructure costs so we can arrive at a unified cost/benefit analysis. Here we continue with our RTI example.

The analysis is based on a system with 60 000 containers. This yields the net result of €235 800 shown in row 1 of Table 4.2. These savings are

Table 4.2 Infrastructure costs for introducing RFID use

RTI Key Figures		Value	Result
1	Savings for the total RTI volume over the total service life of 4 years	60 000	€235 800
2	Annual savings	€235 800 (over 4 years)	€58 950
3	Hardware investment (4 gates and 1 server)	€45 000	€45 000
4	Annual depreciation (4-year linear)	€11 250	€11 250
5	Software investment	€54 000	€54 000
6	Annual depreciation (3-year linear)	€18 000	€18 000
Savings per year			€58 950
Total cost per year			€29 250
Net result per year			€29 700
Net result over four years		€29 700 over 4 years	€118 800

based on a net savings of €3.93 per container over its service life. Row 3 shows realistic order-of-magnitude figures for a system implementation with a scope of four identification points, including hardware and system integration. A customer-specific software solution is also included in row 5. From the net result, it can be seen that the investment in the system is amortized early and that it yields a good bottom-line ROI of approximately 100%.

4.6   Economic Viability Analysis Methods

4.6.1   Requirements for Economic Viability Analysis

We described the advantages of using RFID in very general terms in Chapter 2. In the subsequent chapters, we examined them in increasing detail, since knowledge of the specific processes of the individual parties in the supply chain is necessary for a deeper understanding of the business context. As investment decisions regarding RFID are occupying more prominent positions on enterprise agendas, there is increased interest in examining the potential benefits and using process cost accounting methods to substantiate the results. This requires analysis and quantitative assessment of the distinctions between the main processes and subprocesses shown in Figure 4.25 and the associated activities. It is essential to include every individual process step and every associated activity in the analysis of the current situation and compare them with the target process resulting from using RFID. This requires profound process and RFID expertise, and in some cases it requires extensive time studies.

The variety of activities associated with this can be seen clearly by examining goods receiving subprocesses as an example. In order to

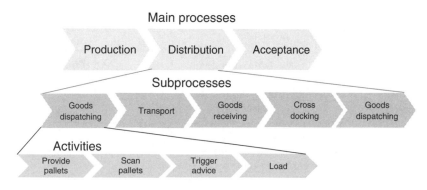

Figure 4.25   Process chain structure

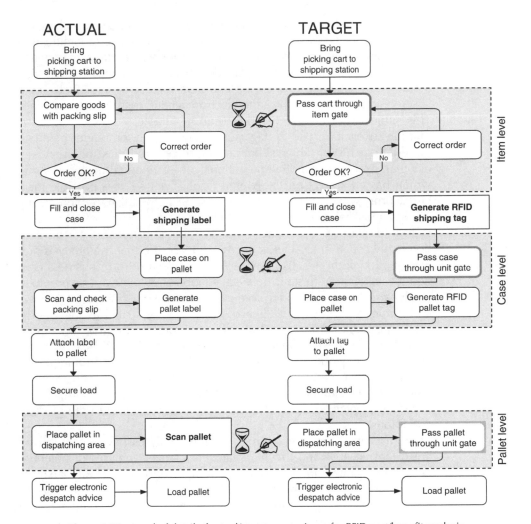

Figure 4.26 Level of detail of actual/target comparisons for RFID cost/benefit analysis

compare the current process without RFID with the target process with RFID as shown in Figure 4.26, it is necessary to describe the activities affected by RFID. After this, the basic values for the comparison must be obtained by making time measurements on the current process. This may not be possible for the target process if no pilot project is available for making measurements. In such a case, preliminary estimates of the changes can be obtained by using a model of a suitable pilot project. In order to obtain reasonable estimates, we recommend that one calls on the services of an expert with practical experience in such projects.

In theory, the model-based estimate could be obtained by making the simple assumption of a 'timeless' activity, since automated identification will occur during the normal process. However, practical experience presents a different picture, since identification becomes increasingly difficult to implement at the case and item levels. A major factor here is screening and distortion of the electromagnetic field by the materials. This subject is discussed in more detail in Chapter 7. The net result is that the recognition rate for tags actually present in the reader field can be less than 100%. Objects that are not recognized must be processed using an exception handling procedure, such as manual postprocessing, which has an impact on the overall result. Even so, it is possible for a target process with a recognition rate of 95% to still be economically advantageous relative to the current process.

In addition to process analysis and model construction, for an economically sound analysis it is necessary to ensure that the analysis and presentation of the results comply with the usual rules of business economics. In addition to static analysis, support should always be provided for dynamic investment analysis methods that take periodic cash inflows and outflows into account. Finally, key parameters such as the amortization period, interest rate, ROI and cash value of the investment can be calculated. Besides this, in the analysis one should bear in mind that there are qualitative factors that must be considered in addition to the quantitative factors. For example, qualitative factors can manifest themselves in the form of increased turnover due to assortment and availability optimization or a reduction in lost sales due to out-of-stock conditions.

Increasingly precise economic viability analyses have been carried out quite recently, and they supplant the blanket analyses of potential benefits generated in past years. Several economic viability analysis models are available, and they can be used to help determine more or less comparable key economic figures such as ROI. Five commercially available tools were subjected to a comparative analysis in April 2006 during a RFID seminar at the German Logistics Academy (DLA). The impressions of approximately 70 attendees were documented using a questionnaire. The economic viability analysis tools listed below formed the subject of the discussion:

- **ROI-Tool** (Seeburger AG, Bretten, Germany)

- **RFID calculator** (GS1, Cologne, with IBM Business Consulting Services)

- **Calculation Chart** (RFID-Kompetenzzentrum, Gera, Germany)

- **RFID Assessment** (Siemens Business Services)

- **rfid-cab** (Logistics Department of the University of Dortmund, in collaboration with integral logistics, a business consultancy based in Dortmund, Germany).

The assessed tools differ in terms of level of detail and thus the complexity of information acquisition. Consequently, these tools are not necessarily suitable for use without a sound process analysis obtained with the assistance of RFID experts. The participants were unanimous in their opinion that the tools should be embedded in comprehensive RFID projects. Nevertheless, they confirmed that all of the discussed tools can be used for fast ad-hoc analyses in order to determine the rough direction of RFID project planning. Information about the individual tools can be requested from their producers.

4.6.2 The rfid-cab Analysis Tool

The rfid-cab analysis tool from the University of Dortmund is described here as an example, since it was generated with the assistance of many participants and has been verified in projects. It was initially developed in a research project of the Department of Logistics (FLog) and subsequently refined in collaboration with business consultancy integral logistics and numerous other enterprises represented in a BVL working group, including Kaufhof, Karstadt, DHL and T-Mobile [Man2006]. The initial target sector was the garment industry, for which ongoing projects such as Kaufhof/Gerry Weber and internal analyses by FLog had generated enough concrete data to enable development of a generally valid RFID economic viability analysis tool [Man2005]. The 'cab' part of the name stands for 'cost and benefit analyser', and it is based on a self-contained program package that supports all relevant activities via suitable graphical user interfaces.

The tool is structured as follows:

- Basic parameter entry – static and dynamic system parameters and general business processes.

- Process analysis – detailed analysis of main processes and subprocesses and their associated activities.

- Evaluation – financial results such as ROI, payment schedules, etc.

- Sensitivity analysis – investigation of the robustness of the obtained results.

The logical steps for determining the results are shown in Figure 4.27.

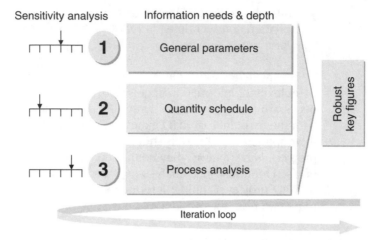

Figure 4.27   Entry screen for general parameters

The quantity schedule is generated first, based on the general parameters, and then the process is defined. The information necessary to obtain robust key figures is described below.

Basic parameters

The basic parameters provide the basic definition of the overall process and the quantity schedule. They consist of:

- General parameters – definition of the general business process, the type of RFID application, and the model assumptions specific to the enterprise:

  - value stages;

  - main processes;

  - identification level (load unit or item);

  - transponder use (disposable or reusable);

  - model assumptions: time frame (periods), depreciation and amortization periods, computational interest rate;

  - transponder costs and returnable-item cycle data: cost per transponder (item or unit-load level).

- Quantity schedule – dynamic material flow movements data:

  - total quantity per period;

  - inventory levels in the individual stages or storage areas.

- Infrastructure costs – RFID hardware and software cost data; used to determine operating costs:
  - RFID hardware;
  - other hardware (PCs, etc.);
  - RFID software;
  - interface adaptation;
  - system integration.

- Labour rates – to determine the labour cost of each activity in the process cost analysis.

- Supplementary potentials – quantifiable and nonquantifiable effects that cannot be derived directly from the process optimization:
  - Goods availability. Is there a possibility of a positive impact on the out-or-stock problem? The extent to which this is possible can only be estimated.
  - Inventory level. To what extent is it possible to reduce safety margins in inventory levels? The magnitude of this depends on the anticipated improvement in logistics quality.
  - Returns. Also affected by improved logistics quality, in particular delivery quality. This aspect is also reflected in increased sales due to improved customer satisfaction.
  - Shrinkage and theft. Increased transparency enables determination of the time of theft, so the perpetrators can be narrowed down and specific measures can be taken. Increased transparency also helps discourage potential thieves, who like to minimize their risk.
  - Increased sales revenue. If the aspect of increased sales revenue is not already reflected in increased goods availability, it can be effected here. Increased sales revenue can also result from improved customer convenience, reducing the volume of counterfeits, and improved customer satisfaction due to better delivery service.

The basic parameters cannot be determined by the RFID project team on its own. All areas of the enterprise that can help improve the quality of the basic data must participate. The sales department is thus just as necessary as the marketing department. The tenability of estimates can easily be revealed by using moderated data acquisition and creating models. Unfounded assumptions regarding supplementary benefits (which are often gladly accepted in order to achieve overall economic viability), such as exaggerated expectations with regard to increased sales, can also be identified quickly – for example, by the marketing department.

Process analysis

Process analysis forms the core of the analysis process, since it allows the effects on the individual activities in the process to be analysed in detail. Figure 4.28 shows the process analysis window.

The 'Prozessauswahl' (process selection) entry pane allows the user to define any desired subprocesses and activities according to the supply structure already specified by the basic parameters. The individual options allow the user to specify the periods with and without RFID and the point of reference (e.g. item level) for the link to the quantities schedule. The structure of the labour cost base is linked to the values acquired under 'Lohnkosten' (labour costs) to enable various wage levels to be represented. Logistics quality factors, such as the error rate in the current process or error reduction from using RFID, can be specified in more detail under 'Fehlervermeidung' (error avoidance). Previously defined processes can also be imported, which means that process chains can be reused once they have been defined. Most of the work for the process

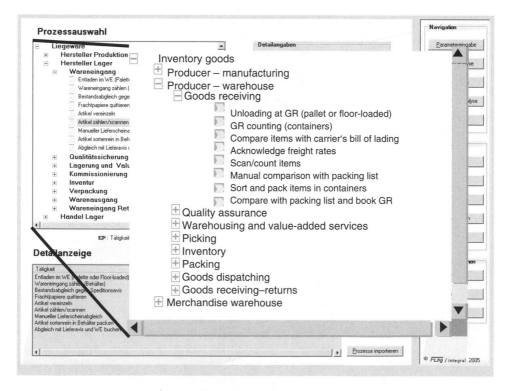

Figure 4.28   Process analysis entry window

analysis consists of physical data input, and the subsequent modelling proceeds quickly.

Analysis results

If the processes are properly represented and the parameters properly defined, the following results can be generated:

- incoming and outgoing cash flows;
- key figures;
- static results (chart);
- amortization interval (chart);
- payment series (chart);
- evolution (diagram);
- error reduction (chart);
- potentials.

Based on the incoming and outgoing cash flows for each period, the key financial figures can be calculated and displayed in a chart. The following key figures can be shown, referenced to the defined levels: net present value, future value, yield, ROI, and dynamic amortization interval. These values are based on an exact determination of all cash inflows and outflows for each period, and they ultimately represent a dynamic investment analysis such as is required for decision support in the enterprise.

The previously mentioned economic viability analysis tools enable more or less similar calculations and analyses. Users must individually decide how they wish to perform their analyses. In any case, it is necessary to attain the level of detail described here, since otherwise it is not possible to obtain reliable results. The stability of the results must also be checked by means of sensitivity analysis.

Sensitivity analysis

The purpose of sensitivity analysis is to assess the sensitivity of a model-based analysis to changes in the underlying assumptions. In our case, this relates specifically to the price evolution of the transponders and the implementation rate or penetration of transponder use. This is comparable to a situation in which a person is considering starting a tour bus company and wishes to estimate the effects of fuel prices and bus occupancy on the operating results with fixed tour prices.

There are many different methods for performing risk estimation for investment decisions. Three sensitivity analysis methods are provided by cab-rfid:

- break-even analysis;
- parameter variation analysis;
- transponder price variation analysis.

The first two of these are standard methods used in investment analysis and are not described further here. This has nothing to do with the importance of the methods, but only with the fact that they are not specific to RFID. This is not the case with transponder price variation. Firstly, the transponder price is a key factor in the economic viability of RFID systems; and, secondly, it is subject to a standard change process. However, price reduction forecasts differ widely, and in hindsight they often turn out to be overly optimistic. In this regard, it is certainly of great interest to determine when a proposed investment would yield a positive return if it does not do so immediately.

4.6.3 Benefits of Detailed Economic Viability Analysis Using Calculation Tools

Our objective in presenting a detailed description of the cab-rfid tool is to demonstrate the need for a thorough economic viability analysis of an envisaged RFID system at an early stage. The previously mentioned tools are suitable for this purpose because they require a broad information base with a high level of detail. This encourages users not only to assess their processes generally, but also to examine individual activities in detail. This will reveal the extent to which they are able to perform an economic viability analysis on their own, or whether they should instead make use of consulting services for this purpose.

We also wish to present a clear picture of the factors involved in initiating a RFID project. In many cases, there will not be a sudden switch to RFID, but staged expansion of RFID use is very important for achieving economic goals. In particular, the break-even point, which means the time when the investment becomes positive, depends on the rate of penetration of transponder use that can be achieved, since the investment in infrastructure components (hardware and software) must be made up front even if only a few objects are fitted with transponders. The analysis tool described here in detail addresses these aspects properly, since it allows penetration processes in the start-up phase, the ratio of disposable to reusable transponders, and the anticipated evolution of transponder prices to be taken into account. With this data, the investment risk can be estimated adequately. The defined price and penetration parameters can also be manipulated during sensitivity analysis in order to evaluate the robustness of the results.

Finally, we can say in summary that a reliable assessment of economic viability, free of any enthusiastic expectations arising from RFID hype, can only be obtained by means of sound, detailed analysis of the potential benefits at the process activity level, combined with a realistic estimate of the rate of implementation.

5

Virtual Identities

5.1 Object Websites

Like other auto-ID technologies, the purpose of RFID is to make real-world data available in real time (i.e. without delay) to IT systems that control and monitor real business processes. Direct transmission of this data is a prerequisite for recognition of the data immediately after the event that triggered data generation and using the data to initiate subsequent actions, such as checking whether a pallet load matches its packing slip.

A serial number stored in the RFID tag of an object is comparable to a social insurance number of a person. Each such number is unique. Just as central data files are maintained for citizens, the idea is to maintain characteristic data of objects on individual websites.[1] This gives each object a virtual identity that can be used as described in this chapter.

The concept of the 'internet of things' is based on the assumption of constantly increasing automation of the link between the Internet, which means the virtual world of information systems, and the real world. This is depicted symbolically in Figure 5.1. The figure shows the real and virtual worlds separated by a funnel that represents this increasing automation, with the distance between the edges of the funnel decreasing over time. This results from continual improvements in data acquisition technologies.

Up to now, deliveries in a logistics chain have usually been checked using paper forms. These forms are collected, and the data and remarks on the forms are then entered into an IT system sometime later by an employee using a keyboard and monitor at a workstation, possibly not until the evening or the next day. There is a more or less significant time lag between the originating event, which is receipt of the delivery, and when the IT system is aware of the event.

[1] The term 'object homepage' is also used.

RFID for the Optimization of Business Processes   Wolf-Ruediger Hansen and Frank Gillert
© 2008 John Wiley & Sons, Ltd

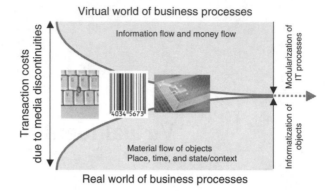

Figure 5.1   Merging of virtual and real worlds with auto-ID technologies

The time lag can be reduced with barcode labels and readers, since the barcode reader transmits the data read from the label directly to the IT system. However, this requires the barcode labels to be oriented so they are visible to the barcode reader, which usually requires manual effort.

This orientation effort is not necessary with RFID tags. No visual contact between the RFID antenna and the RFID tag is necessary. Reading is performed fully automatically using radio signals. Here the only requirement is that the RFID tags must be within the reading range of the reader and must not be obscured by metallic objects or other materials that can interfere with radio signals.

RFID thus closes the gap at the tip of the funnel. The real and virtual worlds are joined together fully automatically without any time lag. The funnel figure in its original form was first published by the Auto-ID Labs in St Gallen [Flei2005]. It also shows that a considerable reduction in data acquisition costs can be achieved by using automated data acquisition (auto-ID), which is also symbolized by the constantly narrowing spout of the funnel. Naturally, this includes not only direct costs for RFID data acquisition, but also savings achieved by improvements in higher-level business processes.

A significant benefit of the internet of things can be expected to be found in improved visibility of logistics paths and transported products. For example, products often appear at the goods receiving area of a warehouse without the information necessary to clearly associate them with a shipment. If this happens, the Universal Product Code (UPC) of the product (such as a Gillette Venus Cartridge in the example of Figure 5.2) can be sent to the object's website on the internet of things. Figure 5.2 shows a possible response to such a query, including the associated case, supplier and recipient (each with its own code) and the shipping data. A suitable action is also recommended – in this case, associating the product with a particular purchase order number. An incorrectly routed

EPCglobal Discovery Station – Gillette's EPCIS Service	
Service Name: Product Authentication	Tag ID sgtin 0 047 400.114008.9780
Product disassociated with Purchase Order	
Dear Wal-Mart,	
Thank you for sending a product authentication query.	
According to our records, the product has been shipped to you.	
Following is the response to your query:	
Customer name	Wal-Mart
Shipped to	Wal-Mart DC 6068R Regular, Sanger, TX
Customer PO no.	1 200 942 202
Container tag ID	sscc: 0 047 400.1300200557
Shipped from DC	Romeoville
Actual shipping date	08/11/2004
Requested arrival date	08/14/2004
Suggested action:	
Please associate the product with the purchase order shown above.	
Additional details:	
Product	Gillette Venus Cartridge
UPC	1 00 47 400 14 008 7
Manufacturing date	07/28/2004
Done. Query timestamp: Sep 21 22:25:13 EDT 2004	

Figure 5.2   Gillette EPCIS service: product identification (source: GS1/EPCglobal)

product can easily get lost if this association is not possible. This leads to shrinkage and lost sales.

What causes such events? For one thing, a case can fall off a pallet or be overlooked during loading. The pallet will thus arrive somewhere with one case too few or an incorrect (not ordered) case in its load. Such situations can be resolved by using object websites. Table 5.1 lists additional questions that can be answered using object websites.

To make it possible to answer such questions, the object websites must provide the following data (among other things):

- Information about design, production, shipping, selling, maintenance and expiry dates.

- Certificates of authenticity and use instructions.

- Data on object movements during delivery, association with pallets or containers used to transport the objects and information on the current location in the supply chain.

For technical products and systems, this also provides support for product life cycle management (PLM).

Table 5.1 Questions that could be answered in the future by the EPCglobal network

Application Area	Query to the EPCglobal Network
Production	We must recall 1 000 parts due to production errors. Where are they located?
Marketing	Product promotion is going very well in the Northeast region. In which regions do we have products in stock that we could shift to the Northeast region?
Retail	We urgently need a follow-up delivery of product XY. Where can we obtain it?
Distribution	The truck has arrived at customer XY, but a case is missing. Where is this case?
Maintenance	When was this part installed? When does it have to be replaced?

5.2   The EPCglobal Network

The EPCglobal network for worldwide data communication was originally conceived by the Auto-ID Center at the Massachusetts Institute of Technology (MIT) in Boston, USA, which also developed the high-level design of the network. Preparations for market introduction of this network were delegated to EPCglobal Inc. (New Jersey, USA, and Brussels, Belgium) in 2003. EPCglobal became a subsidiary of GS1 International Inc. in 2005. GS1 is represented by numerous national subsidiaries throughout the world. In Germany, it is represented by GS1 Germany GmbH in Cologne (formerly known as Centrale für Co-organization, CCG). EPCglobal Inc. (www.EPCglobalinc.org) arose from the merger of EAN International and the Uniform Code Council Inc. (UCC).

We devote a relatively large amount of attention to the EPCglobal network in this book because of the fundamental importance of its systematic structure and international scope. Experts in many market sectors agree that an internationally standardized structure is necessary in order to attain the prescribed goals, such as:

- Increasing the transparency and efficiency of supply chains.
- Increasing transport security (and in particular homeland security).
- Increasing protection against counterfeit products in all markets.
- Using RFID in the information networks of international trade.

- Improving the quality of information flows between enterprises and their trading partners in various global supply chains.

At the same time, the elements of this network are very general and support a wide variety of applications. The details of the operation of the EPCglobal network, which is intended to be the platform for the future internet of things, are not yet clear. Particularly in closed-cycle systems inside enterprises, RFID methods can also be implemented entirely independently of the EPCglobal network.

The trade sector associates the EPCglobal network with wide-ranging visions, which are discussed in the context of international commerce in the Global Commerce Initiative (GCI) [GCI2005]. The architecture of the EPCglobal network is described in illustrative form in a brochure titled *The EPCglobal Architecture Framework* [EPC2503].

Research on the internet of things has been delegated to the Auto-ID Labs. This is a group of research institutions, each of which is associated with a local university. There are seven Auto-ID Labs associated with MIT in Boston (USA), Cambridge University (UK), and the universities of Adelaide (Australia), Keio (Japan), Fudan (China), St Gallen (Switzerland) and Mujiro (South Korea). They were originally funded in part by EPCglobal, but now they are fully funded by industrial partners. Their international distribution reflects the interest of various world market regions in research on the internet of things and their desire to participate in its practical implementation and deployment. The Auto-ID Labs have several mutually complementary focuses. For instance, a research project on developing methods for protection against counterfeiting (anticounterfeiting) is being conducted at St Gallen. The principal focus at Cambridge is on defining tasks in the automotive and aviation industries.

Numerous working groups (action groups) operate under the umbrella of EPCglobal. Their members include experts from enterprises in various market segments and numerous consulting and IT enterprises. EPCglobal is directed by its member enterprises.

In the future internet of things, each object will be assigned its own website, which can be accessed via the EPCglobal network. The basic concept is very centralized. With regard to international trade, it is based on the idea that in the future bilateral EDI transactions between trading partners will be replaced by information about products and their logistical paths available via the EPCglobal network. In this future scenario, a supermarket that receives a pallet with an RFID tag will transmit the EPC of the pallet to the EPCglobal network. The network will consult the associated object website and return the corresponding packing slip information so the supermarket can check the correctness of the delivery and ensure correct handling. In this scenario, there is no need for despatch advice messages transmitted by EDI.

The elements of the EPCglobal network have clear similarities to the Internet, as can be seen from Table 5.2. The EPCglobal network is accessed via EPC Information Services (EPCIS), which are software components designed to act as a gateway between enterprise IT systems and the EPCglobal network.

Table 5.2 Comparison of the structures of the world wide web and the EPCglobal network

World Wide Web	EPCglobal Network
DNS	**ONS**
A central directory that converts Web addresses into IP addresses	A central directory of producers registered with EPC, which converts EPC codes into IP addresses
Website	**Object website**
A virtual location (resource) containing information on a particular subject	A virtual location (resource) containing information about a product
Search engine	**EPC Discovery Services**
A tool for finding websites	A tool for finding EPC information
SSL	**EPC Security Services**
A security standard for websites (Secure Socket Layer)	A tool for secure access depending on assigned privileges

5.2.1 Components of the EPCglobal Network

The purpose of the EPCglobal network is to link producers, trading companies, supermarkets and consumers electronically in the sense of the 'internet of things'. Information about objects is stored in decentralized servers that host object websites. All of this is embedded in the Internet.

The elements and standards of the EPCglobal network are described below. It should be borne in mind that EPCglobal only generates specifications for products and interfaces and does not supply any software products. Products are provided by IT suppliers. Suppliers that are active members of EPC working groups that develop and approve the specifications doubtless have an advantage in this regard.

- **ONS server.** The Object Naming Service (ONS) will be the core component. It will operate in the same way as the well-known Domain Name Service (DNS), which translates URL addresses (Web addresses) into machine-readable IP addresses used to locate websites on the Internet. The ONS will store all EPC numbers issued by GS1 and administer the associated IP addresses of the locations where the object websites are maintained on the Internet. These services will be accessible to users authorized by EPCglobal. An initial version of the ONS has been installed at Verisign Inc. in California. A similar service is provided by Afilias Ltd, Dublin, Ireland.

- **EPC Information Services (EPCIS).** Enterprise IT systems and the EPCglobal network communicate via the EPCIS component, which is a software interface or gateway at the middleware level that is specified by EPCglobal. EPCIS products are marketed by IT suppliers. EPC numbers will be sent to ONS servers via EPCIS in order to obtain the IP addresses of the object websites. The websites can then be accessed via their IP addresses, and the desired product data can be retrieved.

The interactions between the software components for communication with the EPCglobal network are shown in Figure 5.3. In order for this to function properly across enterprise boundaries, there are several standard interfaces that must be taken into account. They are described below. The overall purpose of the network is to acquire, store and exchange data.

As of late 2005, the following standards for an EPC-ready global network had been defined or were in progress:

- **Air Interface Protocol (Gen-2 AIP).** The specification of the air interface protocol (AIP), which is also called 'Generation 2' (Gen-2), governs the properties of the electromagnetic field and the data format for data exchange between readers and tags. It was adopted as ISO standard 18 000-6c in 2006.

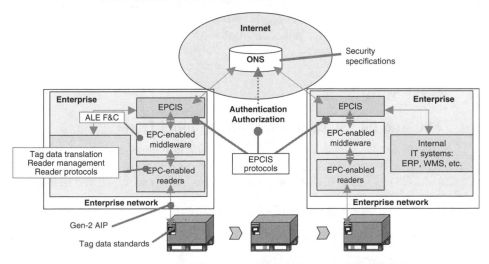

Figure 5.3 Software infrastructure of the EPCglobal network (source: EPCglobal)

- **Reader Protocol.** Describes the data exchange and command structure between EPC-capable middleware and readers.

- **Reader Management Specification.** Standard functions for individual configuration of EPC-capable readers and middleware control of a multireader environment.

- **Tag Data Standard.** This standard describes the memory available in RFID tags that can contain other data in addition to the EPC.

- **Tag Data Translation.** Conversion of EPC data from the tag into an Internet-compatible format, such as XML.

- **Application Level Events (ALE)/Savant.** Specifies how to count, aggregate and interpret EPC information (events) transferred from one or more readers to the middleware. This includes filtering out duplicate read data, which is irrelevant for further processing (filter and collect (F&C) function). This is where the data volume will be reduced to the amount necessary for further processing. The Savant specification is also located at this level. EPCglobal's original intention was to make Savant a product, but this idea was rejected in early 2005, and all that remains is the specification. Savant also includes the interface to enterprise IT systems.

- **ONS Application Layer Interface.** Specification for accessing the EPCglobal network for acquiring information about specific EPC numbers. The interface is at the EPCIS level.

- **Security Specification.** Requirements for secure data exchange between EPCglobal network participants.

The EPCglobal standards provide technical specifications and functional descriptions. Every IT service provider is free to integrate these standards into its products (hardware and software) and make them available to users.

5.3 RFID Standards

5.3.1 EPC and ISO Standards for International Trade

The EPC standards can be used in all enterprises. Specific industry needs are introduced by means of user groups. There are presently GS1/EPCglobal working groups for the market sectors listed below:

- fast-moving consumer goods (FMCG);

- health care and pharmaceuticals;

- transportation and logistics.

Experts from numerous enterprises participate in these working groups to develop specifications and foster standardization.

EPCglobal describes itself as a leader in developing global standards for the Electronic Product Code (EPC) and using RFID technology in global supply chains, particularly in the trade sector. EPCs stored in RFID tags are intended to increasingly supplant EAN codes on barcode labels. However, replacement of barcode labels cannot be expected within the foreseeable future. Instead, the two technologies will be used in complementary fashion (see Table 5.3).

Table 5.3 Commonly used EPC data types (tag data standard) [Kleist2005]

GTIN	Global Trade Item Number: a global EAN/UCC identification number for products and services, with 8, 12, 13 or 14 characters comprising a manufacturer number and a product number
SGTIN	Serialized GTIN: an extension of the GTIN to include a serial number for unique identification of individual objects
SSCC	Serial Shipping Container Code: designation for transportation units with an attribute for fast differentiation between pallets, cases, etc.
GLN	Global Location Number: designates the location of a customer, supplier, warehouse, loading area, hub, subsidiary, etc.
GRAI	Global Returnable Asset Identifier: similar to a GTIN, but used to designate assets or returnable transport items (RTIs) such as drums, gas cylinders, railway cars, trailers, etc.
GIAI	Global Individual Asset Identifier: designates inventory items that remain within an enterprise and are necessary for business activities, such as hospital beds, computers and delivery vehicles
UID	Unique Identification: special numbers used by the US Department of Defense for tracing and tracking assets that can be represented by EPC numbers

The intention is to establish additional working groups according to future requirements. The interests of small and medium-sized enterprises and of countries and regions not yet represented are intended to be included in this standardization process by means of national GS1 member organizations. They are intended to ensure a proper balance of interests and the feasibility of the results in the form of open, global standards.

EPC differs from EAN by having a serial number in addition to the manufacturer and product (class) numbers. The serial number allows individual objects to be identified uniquely. EPCs have a fixed structure as shown by the example in Table 5.4.

Table 5.4 Example of an EPC SGTIN number (length 96 bits) [Harmon2006]

Example: Gillette Venus Cartridge, EPC 0 047 400.114008.9780 (SGTIN structure according to EPCglobal Tag Data Standards Version 1.3, 8 March 2006)		
Field name	Length	Meaning
Header	8	Fixed value: 0 01 1 0 000
Filter value	3	For fast selection of specific logistics objects. For example, '001' designates consumer items
Partition	3	Defines the boundary between the two following fields
Manufacturer number	20–40	Also called 'EPC Manager'. Issued by GS1 and GS1 members: 0 047 400 (as with EAN-13)
Part number (object class)	4–24	Issued by the manufacturer; capacity 1 million object classes: 114 008 (as with EAN-13). The total length of the manufacturer number and part number fields is always 44 bits
Serial number	38	For unique designation of individual parts; capacity per manufacturer 100 billion items: 9 780

The technical basis of RFID systems is described by standards. They specify frequencies, data rates, codings, protocols (including anticollision protocols) and other conditions. Relevant standards are available from the International Organization for Standardization (ISO) and EPCglobal. Of course, it is desirable to have the EPC standards be recognized by ISO as well. For this reason, EPCglobal submitted the Generation 2 (Gen 2) specification for UHF tags to ISO, which issued it as the ISO 18 000-6c standard.

Harmonization of standards for the air interface is crucial for widespread use of EPCs, since complete reading reliability can only be achieved if all readers can recognize all tags that pass through their antenna fields. For the same reason, it is desirable for all tags to work with the same frequency. However, this is not possible for the following reasons:

1. Frequencies are subject to a variety of physical limitations depending on ambient conditions.

2. Different frequencies are available in different parts of the world (see Figure 5.4).

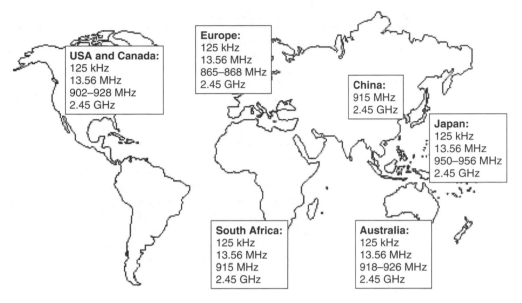

Figure 5.4 Allowed frequencies in world regions (with restrictions)

Consequently, harmonization of RFID tags will not be possible in the foreseeable future.

RFID readers are divided into classes according to various options (see Table 5.5). In order for the options of a particular class to be used, the tags must have corresponding features. EPCglobal initially defined two basic classes (Class 0 and Class 1) for the trade sector, but they were merged into a single class in the Generation 2 standard. Additional classes designate RFID tags with supplementary features. The efforts of EPCglobal are focused primarily on UHF frequencies in the 860–960 MHz band with electromagnetic coupling between tags and readers and a reading range of several metres, which is necessary for identifying objects entering and leaving logistics warehouses.

From the very beginning, some people argued that the 13.56 MHz HF frequency should be used due to its international availability and suitability for item tagging. However, its range is less than 1 m and thus too small for reading pallet tags. On the other hand, the electromagnetic coupling used by UHF tags in the 860–960 MHz band is not suitable for near-field applications. It thus appears that both frequencies will ultimately be used in parallel. This was confirmed by a press release issued by EPCglobal (Brussels) on 4 May 2006, according to which a new working group has been established to extend the Generation 2 standard to HF and thus eliminate the previous restriction to the UHF band.

Harmonization to the 860–960 MHz band is not possible because this band is allocated to mobile telecommunication networks and military uses in various parts of the world. This increases the production cost of

Table 5.5 EPC tag classes for the air interface protocol (AIP)

EPC generations and classes for RFID tags		
Gen. 1, Class 0	Read-only, passive (no internal power source)	860–930 MHz
Gen. 1, Class 1	Write-once, passive	860–960 MHz
Gen. 1, Class 1	[EPC2503] incomplete. Completed in Generation 2	13.56 MHz
Gen. 2, Class 1	Generation 2: a new, uniform specification that eliminates Classes 1 and 2; passive with at least 256 bytes of memory. Adopted as ISO standard 18 000-6c in 2006	860–960 MHz 13.56 MHz (pending)
Class 2	Passive tags with supplementary functions, such as data storage for encryption	860–930 MHz
Class 3	Semipassive tags	860–930 MHz
Class 4	Active tags (with internal power sources)	860–930 MHz
Class 5	Active readers that can communicate with all classes and with each other	860–930 MHz

UHF tags because technical compromises are necessary to produce tags that can be used over the entire UHF band. Either different tags must be used in different areas (e.g. different tags in the USA and Europe), or the tags must operate with less than optimum performance.

Another issue that arose during the early days of RFID tag technology was the allowable antenna power. This was limited to 0.5 W in Europe, which would have restricted tag use. The allowable level in Europe was increased to 2 W in late 2004, which is adequate for industrial requirements. The principal application areas of the various frequency bands are listed in Table 5.6. The *RFID Handbook* [Fink2002] provides additional information about specific aspects and relationships, including electromagnetic factors.

EPCglobal concentrates on standards relevant to object tagging, including transport units in the trade sector. There are also numerous other application areas for RFID-based methods that use different frequencies and standards. Additional information can be found in various ISO standards. The 13.56 MHz frequency used for smart cards is especially prominent in such areas.

As such cards must always be passed close by an RFID reader, the reading distances that can be achieved at this frequency are adequate. In

Table 5.6 Radio frequency bands: properties and applications

Frequency	Properties and Limitations	Applications	Extent
LF: 9–135 kHz	Reading distance less than 1.5 m; also usable in metallic surroundings	Central locking systems for automobiles	Worldwide
HF: 13.56 MHz	Reading distance less than 1.5 m; works on metallic surfaces, but shielding required	Individual items, smart cards, access control, ticketing, e-passports, books	Worldwide
UHF: 860–960 MHz	Reading distance less than 4 m; high reading speed, but works poorly on metallic surfaces, under moist conditions, and at close range	Tagging of pallets and cases; aircraft luggage	Worldwide in various subbands
Microwave: 2.45 GHz	Reading distance approx. 100 m; used in WLAN and WiFi systems	Road toll and container tracking systems	Worldwide

fact, a short reading distance is an advantage because it provides better protection of the cardholder's privacy.

As the technology is constantly evolving, it has become necessary to rework several standards. This has given rise to the specifications in the ISO 18 000 family. These standards also describe the measuring procedures for testing RFID systems, which involve field strength, modulation depth, range and processing time (see Table 5.7).

RFID tags are actually assigned ID (identification) numbers during production that can be used to identify them individually. These are not the same as EPC numbers, which are stored in the tags before they are used with objects. The IDs assigned during production can also be used for unique identification. For example, it is known that clothing retailer Marks & Spencer in England uses such tags for logistical monitoring of its hanging goods. This is a typical example of an application that can be regulated entirely independently of EPC requirements because the entire supply chain is a 'closed cycle' under the control of Marks & Spencer. A similar application used by gardeur ag is described in Chapter 10.

5.3.2 Standards in the Automotive Sector

The automotive sector has international trade structures with complex interrelationships between manufacturers, suppliers and logistics service

Table 5.7 Major ISO standards for RFID

ISO standards		
ISO 14 443 A/B	Proximity cards: air interface and initialization of contactless smart cards; reading distance in the 10 cm range	13.56 MHz
ISO 15 693	Vicinity cards: air interface and initialization of contactless identification cards (smart cards); maximum reading distance 1.5 m	13.56 MHz
New family of ISO standards		
ISO 18 000-1	General air interface specification for internationally accepted frequencies	n/a
ISO 18 000-2	Reading distance of a few centimetres	<135 kHz
ISO 18 000-3	Reading distance up to 1.5 m; successor to ISO 15 693	13.56 MHz
ISO 18 000-4	Reading distance greater than 100 m	2.45 GHz
ISO 18 000-5	Withdrawn	5.8 GHz
ISO 18 000-6	Reading distance more than 4 m, including the version **18 000-6c for EPC Generation 2** (bit 17 = 0)	860–960 MHz
ISO 18 000-7	Reading distance up to 100 m	433 MHz
ISO 18 047-6	RFID conformity tests	860–960 MHz

providers. Like other sectors, it can be viewed from the perspective of the consumers (i.e. motorists) or the perspective of the manufacturers, which now obtain up to 80% of the parts and modules for their vehicles from external suppliers. This makes logistics service providers an important interest group in the supplier network.

This sector differs from other industrial sectors in that new production methods that take this networked supplier structure into account, such as 'just in time' (JIT), are repeatedly developed and introduced in all enterprises in the sector. This means that production plants no longer maintain large parts stocks. Instead, parts are delivered by logistics enterprises directly to assembly-line stations at the right time and installed immediately.

Organization of processes in production cells has also been further refined by application of the 'just in sequence' (JIS) principle. This means that different models and completely different vehicle configurations are assembled in sequence on a single assembly line. The specific assembly step or parts to be installed are displayed in each cell so the vehicle currently passing by on the assembly line will correspond to the customer order at the end of the process.

Automobile manufacturers also use this method to achieve flexibility for assembling different models on individual assembly lines. For instance, BMW assembles its Z4 sports car on the same lines as its X5 SUV. If consumer demand for the Z4 declines, it builds more X5 models, so the assembly line is always fully utilized. To ensure that flexible production yields reliable results, bare car bodies are fitted with RFID transponders with enough memory to store the assembly data for the body concerned. In each assembly cell, the production steps required in the cell are read from the RFID transponder. The transponder is removed at the end of the assembly line and reused repeatedly.

RFID for boosting competitive advantages

JIS organization requires a high degree of visibility of the specific parts needed in each of the production cells. Installation of incorrect parts, which is usually only detected later in the assembly process or not until final inspection, is expensive to correct and dramatically reduces the manufacturer's profit. Here RFID transponders lead to significant quality improvement.

RFID technology is used widely in production control systems in the automotive industry. These systems are monitored by a class of software called 'manufacturing execution systems' (MES). In this case the RFID tags are not attached to parts that are subsequently installed in vehicles, but instead to transport containers and tools. For example, this can be used to let a drilling machine automatically check that it picks up the correct drill bit. For this purpose, an RFID tag is attached to each drill bit and the drilling machine is fitted with an RFID reader. The tag on the drill bit usually contains not only an ID number, but also individual bit data such as the most recent inspection date, shaft length, and so on. With this information, the drilling machine automatically knows how deep to feed the drill or whether it should generate a signal so the drill is sent in for service because it has already been used too often.

RFID processes of this sort are scarcely known to the outside world because they take place inside the production area and are not relevant to logistics. They are also frequently referred to as 'sensor technology' instead of RFID. Another reason that such processes are not discussed openly is that here RFID forms part of an optimized production process that gives the user a competitive advantage and should be concealed from competitors. Market observers that only view RFID from the logistics perspective often overlook this extensive application area.

Optimizing the container management

Another area where RFID is understood to be part of progress is in tracking containers used to transport parts and body panels between various assembly plants. This has been described in numerous publications, such as from Volkswagen. Containers are a vexatious topic in all sectors of industry because they can easily be misrouted. In particular, empty containers often remain in unintended locations instead of being returned promptly for reuse. Consequently, such supply chains often have a rationalization potential of 10% or more. This means that 10% fewer containers would be necessary with optimized container management using RFID. If this involves containers costing a few thousand euros or even ten thousand euros, the savings effect quickly frees up enough capital to pay for an RFID solution. Here again, the number of RFID solutions already being used in industry is larger than is generally known. The basic aspects of container tracking are described in Section 4.5.

RFID methods in the automotive sector become the subject of industry-wide discussion, and thus public discussion, when several enterprises and logistics companies are involved in the container supply chain. This is because in such cases they must agree on RFID standards, in particular part numbers and container numbers, so the flow of goods can be successfully controlled and monitored. Here as well there are clear links to the trade sector. For instance, some replacement parts such as windscreen wipers or tyres are sold via retailers (and associated megastores) that insist on EPCglobal standards and thus EPCs.

This leads to a debate on whether the automotive sector should introduce EPCs as a general standard for both production and trade, since this would eliminate all the problems and make it possible to track all parts as far as installation or the point of sale. However, there is considerable resistance to this suggestion, so the outcome of this debate (which has been going on for years already) is presently unpredictable. The resistance is based on a simple fact: parts numbering schemes are a long-term matter, even inside production companies. This started around 30 years ago with the harmonization of bills of material in manufacturing companies, which was necessary for using comprehensive enterprise resource planning (ERP) software systems. At that time, design engineers quarrelled with production and maintenance experts regarding uniform parts numbering schemes. After these schemes were harmonized inside the company, the next challenge was to agree with external suppliers on uniform numbering schemes. And then companies were merged or integrated into corporate groups, and once again the numbers had to be harmonized. Corporate organizations such as General Motors are still working on harmonization across the entire organization.

In this situation, EPCglobal provokes irritation with its attempt to convince automotive companies to adopt EPCs. Aside from the organizational expense, there are other obstacles here. EPC permits only numerical codes and does not allow alphanumerical codes, and it has a restricted code length. This means that existing alphanumeric codes used in the automotive sectors cannot simply be copied to the EPC format. A knowledgeable observer may conclude from this situation that EPC will not quickly assume a significant role in the automotive sector. The result will probably be that manufacturers will attach supplementary RFID tags with EPCs to their retail products, even if this generates additional costs. By contrast, the cost of an industry-wide switch to EPC would be enormous and thus prohibitive.

RFID activities at Odette

The European automotive industry association Odette, which is based in London, UK (www.odette.org), is more significant than EPCglobal with regard to global standardization in the automotive sector. Odette is a nonprofit organization 'run by automotive people'. Its mission is to develop tools and recommendations that improve the flow of goods, services, product data and business information along the entire supply chain and during the entire product life cycle. Its principal areas of activity are e-business communication, logistics management, and engineering data exchange.

As Odette states on its website (November 2007):

> The necessity for parts identification and traceability arises from the requirement to be able to identify defective parts in the field and from the human and financial need to be able to reduce any risk of damage after the point of sale. Up to now each company has regulated parts identification and traceability individually. So far so good – but at the interface between companies it becomes more difficult. There is no clear agreement on the delimitation accuracy required for parts and their components, who stores which process/quality data relative to which references, or which references are to be communicated to the customer and linked to the customer's product. Also, different coding and presentation formats increase the technical effort at the supplier side and make data capture difficult.

There are various activities within the Odette community under way to overcome these shortcomings, including the RFID Project Group comprising the subgroups container management and vehicle identification. In Germany, the VDA (Verband Deutscher Automobilhersteller; www.VDA.de), with headquarters in Frankfurt am Main, is a member of Odette as a national industry association. Other members include manufacturers and suppliers such as Bosch, Audi and ZF. The following projects are presently underway at VDA:

- Standardization of vehicle shipping information for RFID use. This project focuses on the logistics process from the end of the manufacturer's assembly line to delivery to the dealer and the potential benefits of using RFID tags, in particular in combination with plans for a standardized vehicle shipping label.

- RFID for tracking parts and modules in the automotive industry.

The Vehicle Identification Number (VIN), which was standardized in the ISO 3779 international standard in 1983, is an essential data element for all planned projects, including those of Odette.

VDA issued Recommendation 5501, which deals with RFID for container management in the supply chain, in November 2006. This recommendation forms the working basis for the Odette container management subgroup. The recommendation addresses optimization of container management with the following objectives:

- standardizing the use of RFID components in container management;

- standardizing the data to be stored in the transponders;

- requirements for the accompanying EDI messages.

This is based on use of the ISO 18 000-6 and 18 000-6c (EPC Gen2) standards. EPC and the Dun & Bradstreet DUNS scheme are both being considered as numbering schemes. Here again we see the problem that manufacturer number schemes such as DUNS that are already being used in this industry cannot easily be replaced by EPC. The VIN scheme is also incompatible with EPC because it uses alphanumeric codes.

A paper by Craig Harmon, 'The necessity for a uniform organisation of user memory in RFID' [Harmon2006], can be recommended for more detailed study of the data format of RFID transponders. It also clearly shows that the 'EPC standard' (ISO 18 000-6c) is simply a variant of ISO 18 000-6. A '1' in bit position 17 of the RFID transponder data format indicates that EPC is being used. If this bit is set to 0, any other desired code can be used. More detailed specifications can be expected from the ongoing activities of the Odette organization.

The specific role of logistics services providers

With regard to the logistics process of international delivery of motor vehicles, it must be borne in mind that this is not a pure supply process. In many cases, logistics companies also perform final assembly activities as part of the logistics process. Here there are already numerous RFID solutions, which are also reported in the press. For instance, BMW uses a real-time locating system (RTLS) from Wherenet in its car parks. This system works in the same way as a WLAN. Each car is fitted with a RFID transponder having a reading range of more than 100 m. Each car in the car park can be located precisely, even in a multistorey car park, by using several antennas to triangulate the RFID signals. This speeds up the process of loading vehicles onto lorries and ships and reduces misrouting of vehicles. Systems of this sort are also used at the loading stations of large maritime ports. The RFID transponders are removed when the vehicles are loaded, after which they are reused. Volkswagen uses a similar transponder-based system from Identec at its Wolfsburg site. An extensive example of a locating system used by a car dealer, which combines RFID and GPS, is described in Chapter 10.

Improving traceability

There is a research project funded by the German government, with participation by research institutions and industrial firms, with the objective of improving traceability [Laendm2007]. According to the Germany Federal Motor Transport Authority (Kraftfahrt-Bundesamt), 167 recall campaigns were conducted in Germany in 2006 – more then ever before. They involved more than a million vehicles. These recall campaigns are very damaging to car manufacturers, not only because the cost of correcting such quality shortcomings can easily be a seven-digit figure in euros, but also because the associated damage to the manufacturer's image is nearly inestimable.

The Laendmarks consortium intends to develop a comprehensive system that can be used to trace quality problems in individual components back along the global supply chain to their respective manufacturers. The consortium is headed by Keiper, one of the largest suppliers of seat components and structures to the automotive and aircraft industries. Daimler and Volkswagen are also involved on the manufacturer side, as well as the chair of mechanical engineering and informatics at Ruhr University Bochum in Germany.

The great importance of product traceability is also recognized outside the automotive industry. The food and pharmaceutical industries are making intensive efforts to fashion transparent supply chains. Consequently, the Laendmarks project intends to develop models and processes that are as versatile as possible so they can be used in other sectors as well.

5.3.3   Standards in the Aviation Sector

The common interests of aircraft manufacturers and airline companies are looked after by the Air Transport Association (ATA) and its international affiliate, the IATA. For many years, discussions regarding standards for electronic product labelling with RFID have been conducted in these organizations. Boeing and Airbus also organized a series of international conferences for their suppliers in order to promote the use of RFID tags and electronic codes. Regulatory bodies in the aviation sector, such as the Federal Aviation Administration (FAA) in the USA and the Luftfahrtbundesamt (LBA) in Germany, are also involved in these activities.

The applicable ATA specification is Spec 2000, which is not compatible with the GS1 EPC specification. However, the airlines have now recognized that they cannot simply isolate themselves from the trade sector. For instance, aircraft catering involves items originating from the trade sector, for which EPC will be used. Goods transported by aircraft as part of logistics chains in the trade sector will also bear EPC tags, while aircraft parts will be marked with ATA codes.

This ambiguity cannot be eliminated in the short term by harmonization, so the aviation sector is taking a dual-path approach. It accepts EPC in logistics areas, but it uses Spec 2000 for aircraft parts, particularly for maintenance, repair and overhaul (MRO) (see Figure 5.5). However, the long-term goal is to achieve harmonization of the standards, as indicated by the arrows pointing towards EPC in the bottom right portion of the figure.

It should also be noted that in the aviation sector, electronic systems used in aircraft must be approved by aviation regulatory bodies. The provisions with regard to technical parts with RFID tags that are intended

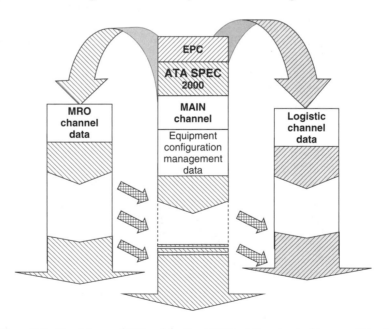

Figure 5.5   Coexistence of EPC and the Spec 2000 aviation standard (source: Airbus)

to remain on the parts during aircraft operation are very restrictive. Approval of tags on transported goods is less stringent. In the summer of 2005, the FAA finally approved use of passive RFID tags on transported goods for operation at the frequencies of 13.56 MHz, 915 MHz and 2.45 GHz. This approval is subject to the condition that the tags must not be activated during flight, which is actually self-evident.

In the autumn of 2005 Boeing announced that it intended to fit around 2000 parts of the new 787 Dreamliner aircraft with passive UHF RFID tags with 64 kilobits of memory, and that these tags would serve to identify the parts and record their maintenance history [OCon2006]. They were supposed to use the EPC Gen 2 standard. However, the rumour is that this plan has not yet materialized and that even procurement of 64-Kbit RFID tags is not certain.

At Airbus, plans for using RFID technology were a closely guarded secret until the company invited *RFID Journal* to visit the A380 final assembly plant in Hamburg, Germany, in the summer of 2007 for a first-hand view of some of Airbus's RFID deployments and pilot projects [Wass2007]. According to the report in the magazine,

> The first phase, which is in deployment, involves tracking supplies and reusable containers with their contents through the supply chain to assembly plants. The second phase, which covers manufacturing and assembly, includes a pilot to track large components, such as wings and fuselage sections, as they are transported throughout Airbus's global supply chain. The third phase, which is set to begin pilots early in 2008, looks at in-service support processes.

Airbus is trying to encourage its suppliers and customers to consider the benefits of RFID by demonstrating how it is using the technology to achieve results in its own business. For example, RFID helps ensure that the 800 parts containers needed to build each cabin get to the right place at the right time. However, as already mentioned it will take some time before RFID emerges from the pilot stage and finds widespread use in the company.

Lufthansa German Airlines is presently using RFID tags on aircraft seats (see the case study in Chapter 10). This application does not cover all airplane seats, but is instead limited to 'quick change' seats, which means seats in business and economy class that are reconfigured at the airport between flights to match the passenger booking situation.

Several pilot projects for tracking and tracing air passenger baggage with RFID have been also been conducted successfully. A problem here is that the involved parties, in particular the airlines and airport managers, have not yet managed to agree on a standard for RFID tags or a way to share the costs of RFID tags on baggage. Even if they decided to use RFID, it would take several years to introduce it because replacing barcodes with RFID would require fitting airport conveyor systems with RFID readers instead of barcode scanners, which would involve considerable expenditures of time and money and would have to be coordinated with depreciation cycles.

One new product that could help stimulate the use of RFID in cross-enterprise logistics chains is SITA's Auto ID Managed Services, which have been presented at several conferences including the 'Dortmund Meetings' in September 2007 in Dortmund, Germany. These services are described in more detail in Section 6.3.2.

5.3.4   Standards in the Pharmaceutical and Health Care Sectors

The most important driver for RFID use in the pharmaceutical industry is protection against counterfeit medicines (anticounterfeiting). Industry statistics show that the impact of counterfeits extends beyond the loss of income suffered by manufacturers of genuine products. A much more serious aspect is that 95% of counterfeit medicines have little or no therapeutic value because they lack important ingredients. This represents a major threat to public health. For example, counterfeits forced US manufacturer Pfizer to recall 19 million Lipitor tablets, including the counterfeit tablets, at a cost of US$55 million.

Furthermore, in the case of counterfeit medicines it is impossible to ensure compliance with required transportation conditions such as a specified temperature range.

The US Food and Drug Administration (FDA) therefore stipulates that in the future all medicines must have serial numbers and that electronic histories (e-pedigrees) must be maintained for them. The intention of this is to make the entire supply chain transparent in the pharmaceutical industry. The FDA recommends attaching RFID tags to medicine packaging for this purpose, but this goal could also be achieved by using two-dimensional barcodes, such as Data Matrix. The first legislative provisions for electronic pedigrees of pharmaceutical products have already been enacted in Florida and California (USA), and they entered into force in 2007.

There are also some initial statutory regulations in Europe. For instance, Italy stipulates that in the future, medicines must have electronic tags with serial numbers and all transactions in the pharmaceutical market must be registered with a central supervisory body.

The code use situation in Europe is difficult because there are several different code schemes. The Health Industry Barcode (HIBC) is used in the Netherlands, while the Pharmazentralnummer (PZN) ('central pharmaceutical number') for medicines has become established in Germany. PZN numbers are issued by IFA GmbH, which in Germany plays the same role in this sector as GS1 in the trade sector. The PZN scheme forms the basis for account settlement with health insurance plans, so it cannot simply be replaced by a different code.

There is also the European Health Industry Business Communication Council (EHIBCC), which issues HIBC codes. This association is also unwilling to simply relinquish its activities in favour of GS1 (see the Berlecon Report [Ber2005]).

5.3.5   Standards in the Military Sector (US DoD)

Logistics expenses in the military sector are enormous and international. From studies on using RFID tags, the US Department of Defense (DoD) has already concluded that savings amounting to billions of dollars could be achieved with RFID tags. Consequently, from 2007 onwards all 60 000 suppliers are required to attach RFID tags to their shipments.

Since many items supplied to the military are trade products that will be marked with GS1 EPC codes either directly or on their packaging, the DoD is working to achieve harmonization between these codes and the following military codes:

- CAGE: Commercial and Government Entity – used on packing slips.
- DODAAC: DoD Activity Address Code – used for logistics purposes.

For future shipments, the DoD will allow suppliers to store these codes or EPCs in the tags. RFID readers can detect the difference from special information in the EPC header. This also enables military codes to be stored in ONS servers for looking up their websites on the internet of things.

5.3.6   IEEE Standard

The Institute of Electrical and Electronics Engineers (IEEE) is a renowned association that generates a large number of standards. The IEEE also has an RFID standard for a 64-bit code, which differs from the EPC standard in that it has only two fields:

- manufacturer: 24 bits;
- extension identifier: 40 bits.

Unlike the GS1 EPC, this code does not have a header and thus cannot be classified into different application areas. It also does not make a distinction between item numbers and serial numbers, so it appears to be less suitable for typical uses in the logistics chains of the trade sector.

From IEEE documents, it appears that it is intended to be used for labelling electronic parts (http://standards.ieee.org).

5.4   GS1 EANCOM EDI Standard

Electronic data interchange (EDI) methods are intended to foster inter-national electronic transmission of commercial documents between enterprises that trade with each other or maintain other types of business relationships associated with delivery and settlement. Typical examples of EDI documents are despatch advice messages, receiving advice messages, inventory level messages and invoices. There are many different types of EDI methods in different industrial sectors. Use of EDI is a prerequisite for widespread use of open RFID systems that extend across enterprise boundaries. For example, a pallet with an RFID tag that arrives at a customer warehouse can be electronically compared with the despatch advice and checked for correctness. This speeds up pallet receiving, and it makes handling of complaints and returns faster and more businesslike. As a consequence, invoicing is faster and the cash situation of the supplier is improved.

GS1 and EANCOM have jointly generated a standard for EDI methods [GS12005a], which is accepted in the trade sector but not yet used to a sufficient extent. The standard defines the contents of the various commercial documents and the formats of the transmitted records. This enables recipients of EDI documents (or their IT systems) to correctly interpret the commercial contents, which are simply strings of bits at the electronic level.

An important consideration is that all shipping units, such as pallets, must have unique numbers, which in this case are Serial Shipping Container Code (SSCC) numbers (in German: Nummer der Verpackungseinheit, NVE). In the future, these codes will be assigned uniquely to shipping units, and they will be stored in electronic form in RFID tags and printed on the usual transport labels in the form of text and EAN 128 barcodes (Figure 5.6). The RFID tag will be located either invisibly on the back of the label or in a side pocket that can be opened to improve the effectiveness of the RFID antenna field. In addition to the SSCC, other logistically relevant information can be shown on the label, such as the batch number, best-before date, and weight.

When a shipment is received, the SSCC and other data is read from the barcode or by an RFID reader if an RFID tag is present and then compared with a previously sent EDI message (i.e. a despatch advice message (DESADV).

The EDI messages for a typical delivery process, as specified by EANCOM, are described in Figure 5.7. The supplier, recipient, logistics warehouse and logistics service provider are involved in this process. The logistics centre in the figure represents other distribution centres and intermediate warehouses, which can further expand the logistics chain:

- DESADV: despatch advice

- HANMOV: handling and movement

- INVPRT: inventory report

- IFTMIN: transport/forwarding instruction

Figure 5.6 EAN 128 shipping label (source: GS1)

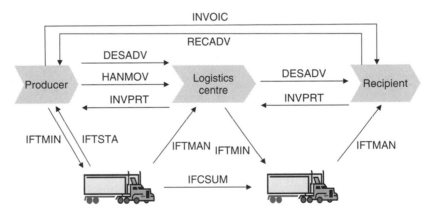

Figure 5.7 EANCOM message types (source: GS1)

- IFTSTA: multimodal transport status
- IFTMAN: freight forwarder arrival notice
- IFCSUM: forwarding and consolidation summary
- INVOIC: invoice.

The fact that establishing this sort of EDI message traffic is a prerequisite for obtaining any benefit from using RFID methods is often overlooked.

Received shipments can only be associated with other information if an electronic despatch advice is available. Likewise, it is especially easy for freight forwarders to identify the shipping containers they have been ordered to transport if they have received electronic notification in advance.

6

IT Architectures and Services

6.1 RFID and Higher-Level IT Architectures

6.1.1 Introduction

The first thing that comes to mind with RFID methods is RFID tags and antennas that communicate with the tags via radio signals. Implementing the reading or identification processes is apparently very simple, but what happens with the data that is read from the tags into the memories of the readers controlling the antennas? Where does the data go next? How can the data be used to provide reliable information to enterprise information systems? To achieve this, the peripheral RFID components must be integrated into an overall system concept. They must be part of the IT architecture of the enterprise that wishes to utilize RFID methods. This is why considering the system architecture from the IT perspective is worthwhile in the context of RFID.

The term 'architecture' originates from building construction and designates the inner and outer structure of a building. An architecture can only be implemented if three factors are in harmony: the external form (aesthetics), the engineering design, and the physical construction. If an architect ignores this interrelationship, his or her design cannot be implemented. This can be illustrated by a well-known example: the Sydney Opera House, with its sailboat silhouette, could not be built in 1965 as originally designed because the architectural concept was too sophisticated. Further advances in building engineering were necessary to allow its sophisticated shapes to be cast in concrete.

The term 'architecture' is used in many disciplines, including informatics. It is relatively easy to describe the interactions among hardware components, because they are tangible items such as circuit boards, enclosures and cables. The software of an IT architecture is equally important, but it is virtual instead of tangible. It also consists of modules, interfaces and communication links that must be technically compatible.

RFID for the Optimization of Business Processes Wolf-Ruediger Hansen and Frank Gillert
© 2008 John Wiley & Sons, Ltd

This is why there are job designations such as 'software architect' and 'software engineer'. A software architect is responsible for designing software structures, while a software engineer is responsible for implementing them. There are also other specialists who do the programming.

The task of the software architect is to ensure that the requirements of business processes are correctly mapped into the software, since software systems are actually virtual representations of these processes and the organizational infrastructures of enterprises. The term 'enterprise model' is also used to designate what is represented by the software and databases.

Software development was originally an art and had little in common with engineering disciplines. People simply wrote programs without worrying about imbedding individual software applications in the process structures of an enterprise. The complexity of the software increased with the complexity of the demands placed on it, and quite a few software systems collapsed sooner or later due to internal structural deficiencies, just as a building can collapse if its foundation or lower floors are not designed to bear the load on top of them. This makes software increasingly difficult to maintain; maintenance costs increase to the point that maintaining the functionality of the software is no longer affordable.

In order to avoid this situation, software must be built according to comprehensive concepts that ensure a robust combination of software modules, hardware components and communication networks. In addition, it must be possible to extend the functionality of the software. Software architecture is thus a crucial discipline in informatics, and auto-ID or RFID systems must be embedded harmoniously in software systems. Planners of auto-ID systems must bear this in mind, as otherwise they will implement components on the periphery that cannot be integrated into the overall IT system.

As the term 'architecture' is often discussed on the basis of widely differing backgrounds and experience, here we have chosen to introduce the topic on the basis of an evolutionary process that began with mainframe computers and IBM SNA systems. Starting from this, we describe the development of networks based on the seven-layer OSI model and the Internet. In parallel with this trend, middleware became increasingly important and ultimately emerged as the key infrastructure for the 'real-time enterprise' (RTE) of the future, or for 'real-world awareness' (RWA) in the form of a service-oriented architecture (SOA). We thus devote a separate section of this chapter to the RTE architecture.

The technical aspect of the service concept in SOA is complemented by the business aspect, which is found in the term 'on-demand services'. IT systems can be operated as on-demand services inside enterprises or by external suppliers. The latter situation is also called 'service outsourcing'. On-demand methods are especially important with regard to auto-ID systems because they are commonly used in logistics chains where the data they capture must be transferred to the IT systems of several enterprises. In many cases, this can be realized better by an external service provider than by the joint efforts of the enterprises concerned. Consequently, this chapter concludes with a section describing on-demand services, which are also called managed services.

The subject of agent technology is included in this chapter because it represents a new paradigm that is emerging in the market and will play a major role in the RFID domain. Up to now, data read using RFID have always been transmitted to centralized background IT systems for use in processes supported by these systems. In the future, peripheral objects will also communicate directly with each other and perform processing and decision-making processes on their own using built-in processors or processors allocated to them. Agent technology is thus an innovative extension of the IT architecture at the periphery of enterprise-wide IT systems.

6.1.2   Development of Distributed IT Architectures

IBM SNA network architecture

One of the first architectures for commercial information systems with a significant software background was IBM's System Network Architecture (SNA), which was developed in the mid-1970s. SNA was the first architecture that described the functional structures of communication systems linking several computers and terminals. Until then, the term 'architecture' had not been used for such systems because communication links were only implemented as point-to-point connections via dedicated or dial-up lines. In addition, terminals were always connected directly to application systems. If someone wanted to use a second application from the same location, a new line was necessary. SNA introduced flexibility into this situation, so communication lines could be assigned dynamically under software control or shared between applications.

SNA was a very progressive step in making networks systematic and flexible, and it was adopted worldwide by nearly all large enterprises. Other producers of IT systems that wished to integrate their products into SNA systems were compelled to align their products to this model and support SNA interfaces. Digital Equipment Corporation (DEC) made a similar attempt to define a comprehensive network architecture under the name 'DECnet'. However, it could not compete with SNA, and SNA became the de facto world standard.

At the same time, the nonproprietary seven-level OSI model was developed by the International Organization for Standardization (ISO) (see Table 6.1). OSI stands for 'open systems interconnection'. As with other architectures, here the aim is to classify functional components and communication interfaces into hierarchical layers and specify protocols for communication between the layers. This enables every producer to

Table 6.1 The OSI seven-layer model, which forms the basis for open network architectures

Level	Level Designation	Services and Standards
7	Application software	HTTP, DNS, HTML, XML
6	Presentation	EBCDIC, ASCII
5	Session control Interhost connection	RPC, DCOM, CORBA
4	Data transport End-to-end control	TCP
3	Network connection Path control	IP
2	Physical data transport Bit streams	Ethernet, Token Ring, WiFi, GSM, GPRS
1	Physical medium Cable	Wires, radio signals, glass fibre, RS 232, ISDN, DSL

develop software components that map the functions of a particular layer and can communicate with components from other producers in the layers above and below. This yields an open, nonproprietary network structure. Nevertheless, there are many similarities between SNA and the OSI model.

The next step was taken by IBM, which introduced an architecture for the top layer (application software) called System Application Architecture (SAA), which also described the interface between application software and presentation in graphic user interfaces. This was done in a highly proprietary manner in keeping with IBM's marketing policy, which always assigned a decisive role to central mainframe computers. This was made especially clear by the requirement for a central enterprise repository, which was a database intended to contain all important application and control data. However, central repositories of this sort were never implemented in practice. This centralized, deterministic architecture was a much too complex approach and was doomed to extinction like the dinosaurs.

IBM's SAA also experienced increasing competition from the Microsoft Windows world and Unix systems, which conquered the decentralized departmental IT domains. As a consequence, producers of ERP systems (such as SAP) that had initially based their software systems on centralized models began to produce modular systems, including Unix-based systems. This trend was also called 'downsizing'. It transformed IBM-style mainframe computers into large servers, and they relinquished their dominant hierarchical position in favour of a cooperative model of distributed, networked systems called a 'client/server architecture'. Finally, increasing reliability of the Internet as a medium for robust network connections led to a widespread decline in the use of SNA.

Client/server architecture

Development of the client/server architecture was a major factor in the demise of the centralized IBM architecture (see Figure 6.1). The client/server architecture began as an arrangement of application systems distributed over several servers and communicating with client computers at user workstations.

The next refinement was to generalize the arrangement by allowing every process on each computer to use the services of the other computers. This freed the client concept from its direct link to the hardware. Every form of communication between different programs can be regarded as a client/server relationship, regardless of the hardware platform or platforms that host the programs concerned. This arrangement was finally perfected by object-oriented technology, which is described in more detail in Section 6.1.4. This definitively freed network architectures from the yoke of hierarchical structures and gave them a horizontal structure, with no computer in the network having a predefined dominant role.

The triumphal march of the Internet and open systems

For a long time, SNA communication networks formed the basic infrastructure for global communication due to their high degree of functionality and security. Now they have been replaced in all sectors by Internet-based virtual private network (VPN) links with the same level of functionality and security. These links act as pipelines for data transport (including globally if necessary), and they are effectively screened from other Internet users.

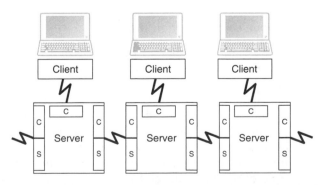

Figure 6.1   A client/server architecture with client workstations linked to servers

The Internet is based on standards, such as the TCP/IP Internet protocol, that are no longer dominated by individual producers. The Transmission Control Protocol (TCP) provides session control at layer 4 of the OSI model, while the Internet Protocol (IP) ensures correct data transfer at the individual link interfaces (OSI layer 3). The layer structure of the Internet does not map exactly into the OSI model, but the differences can be neglected here. They arise from the fact that the Internet has developed pragmatically into a de facto standard, while the OSI model is a theoretical concept.

XML

In the course of the expansion of the Internet, another standard called Extended Markup Language (XML) was developed for transferring data between application systems. This standard was issued by the W3C consortium. The basic idea of XML is to separate data from its visual representation in order to support content transfer with various, selectable forms of presentation. XML provides a syntax for formulating commands and data that can be transmitted via the Internet to any desired communication partner. They can be read visually by humans or computationally by machines. The syntax is located at a logical level and is independent of the hardware. XML messages are generated at layer 7 of the OSI model and utilize the services of the underlying functional layers for data transmission.

XML defines rules for structuring documents. The structure of an XML document is described by a schematic language. Two well-known versions are Document Type Definition (DTD) and XML schemes. DTDs specify the formats of XML documents that are used to describe media, such as articles and books. No distinction is made between text and numbers in a DTD. An XML scheme is an extension of a DTD. It can be used to restrict the structure of an XML document to elementary data types, such as numbers, dates or text.

Mobile communication with GSM

In the mobile telephone sector, GSM is a highly successful standard for dial-up connections, and it allows mobile telephony to be used worldwide for data transmission and voice communication. With the advent of GPRS and UMTS, it is now possible to establish VPN links in the GSM network that provide continual contact between GSM communication devices and application servers. For example, this sort of link is used when GPS satellite navigation devices operate with moveable objects such as containers in order to monitor logistics routes. Position data and other data is transmitted using GSM, GPRS or UMTS.

Extending the life cycle of ERP systems

Increased flexibility of networks and software architectures has allowed modern software to be modularized to the point that individual functional blocks can be distributed dynamically over the network in order to best fulfil users' performance requirements. However, there are some critical exceptions here: most software systems used in the financial services and banking sector and for supply chain management (SCM), warehouse management systems (WMS) and other operational enterprise tasks are still based on the centralized, monolithic mainframe model, even if they currently run on modern platforms such as Unix. They are also referred to as the ERP (enterprise resource planning) class of systems or 'legacy' systems. This emphasizes the fact that the internal structures of these systems generally do no correspond to the current standard of modern, flexible software architectures. This should not be understood as a criticism, since there are economic reasons for this situation. Such software has been repeatedly refined and enhanced by its developers and users over the course of many years of operational use. This represents enormous investments, and reasonable service life is necessary to recover these investments. However, ERP users should realize that there will come a time when these systems have become so inflexible and their maintenance has become so expensive that procuring a new system to support their enterprise processes will be a more economical solution.

6.1.3   EAI: Enterprise Application Integration and Middleware

A wide variety of internal system services for communication between application systems are grouped in a special software layer called middleware. Middleware effectively provides the 'courier services' in networks. The term 'broker services' is also used here, since these links between changing platforms must operate in the same way as a securities broker in a financial services business. For example, public services are implemented using the Common Object Request Broker Architecture (CORBA).

When these broker services are used to access resources available on the Internet via their Web addresses (uniform resource locators, URLs), they are called Web services. This is reflected in product names, such as IBM WebSphere, BEA WebLogic, and WebMethods from the like-named company. Web services form the basis for SOA (see Section 6.1.4).

An essential feature of middleware is that it enables the application software in layer 7 of the OSI model to request all necessary services from other application systems via the network. It thus frees the application software from the need to deal with the technical details of the necessary network connections. These details are configured as necessary by the middleware in order to establish the connections.

The importance of middleware has gradually increased to the point that enterprises want to have seamless IT support for business processes

extending throughout the entire enterprise or across enterprise boundaries, such as in logistics chains. The ability to achieve this sort of IT
integration has become a critical factor for controlling such processes,
and thus for corporate competitiveness.

As a result, the degree of coupling between the information systems
of various departments in an enterprise, and of various enterprises with
mutual business relationships, has increased steadily. In a commercial
world based on division of labour, these relationships extend around
the world, and the importance of middleware for proper operation
of networked IT systems has thus increased proportionally. Due to the
strategic importance of middleware, it is now more commonly designated
in terms of the task this system layer is intended to master, which is
enterprise application integration (EAI).

Integration has become at least as important as the functionality
of enterprise applications. This has an impact on ERP software. In
early 2004, SAP introduced a new product family for EAI that goes
by the name NetWeaver. The name suggests its intended function, which
is to 'interweave' different IT systems across enterprise boundaries.

Transferring this integration function to the middleware layer has the
enormous advantage that ERP systems can continue to exist and be
refined in the form of enterprise servers. They can also be rehosted to a
different operating system platform without any impact on the network.
For instance, many ERP systems that ran under proprietary operating
systems at the beginning of their life cycles currently run on Windows or
open Unix platforms. The middleware provides the interfaces necessary
for communication with these systems. The client/server model thus
remains intact (see Figure 6.2).

This separation of communication control from applications also forms
the basis for flexible extension of enterprise IT systems to adapt them to
ever-changing forms of business processes with potentially global scope.

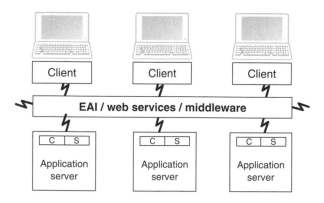

Figure 6.2 A client/server architecture in which middleware provides system services and
enables flexible links between clients and servers

6.1.4 Service-Oriented Architecture (SOA)

All IT producers that wish to play a significant role in the enterprise IT sector have adopted this model, and they are all working to refine it. Generally speaking, the software used in RFID readers and the servers to which they are connected supports this architecture.

Figure 6.3 shows the historical progression of IT architectures from the centralized mainframe orientation to the current SOA model. Here it is also necessary to consider the phenomenon of events, which are fundamentally important for auto-ID applications (see Section 6.1.5).

In the 1980s, client/server technology led to the integration of personal computers and departmental computers to form extensive networks in which application systems could be modularized and distributed. Object-oriented technology was the first approach to consistently extend this scheme to the software component (object) level. Another significant platform for this is CORBA, which was developed by the industrial consortium OMG. CORBA includes a tool called 'remote procedure call' (RPC), which enables an object to invoke a link to another object. CORBA is described further in Section 6.2 in connection with software agents, since they are typical examples of software objects.

SOA has transformed this model into a generic infrastructure for enterprise IT systems. The network services for this are implemented by the EAI functional layer, which is the middleware.

The SOA service model

A SOA is composed of three elements: consumers, suppliers and a directory (see Figure 6.4). Suppliers publish descriptions of their services to the directory. Consumers can search the directory, select a desired

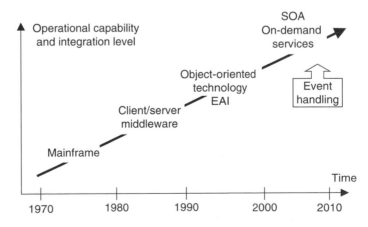

Figure 6.3 Historical development of the service-oriented architecture

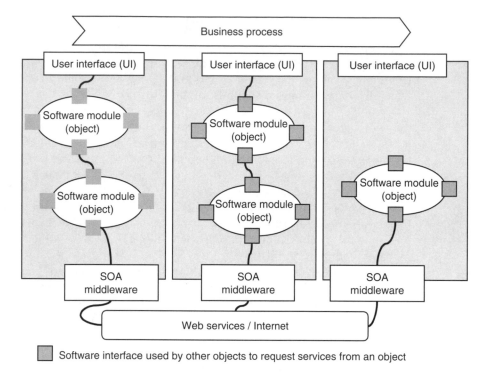

Software interface used by other objects to request services from an object

Figure 6.4   SOA with flexible provision of software services

service and invoke it remotely. The concept of remote method invocation is similar to the remote procedure call in CORBA.

The dominant position of the mainframe computer came to an end during the development of the SOA model. The dominance of ERP suppliers was also tempered, and producers were forced to constantly redefine their positions during the course of this development. A clear expression of this metamorphosis can be seen in IBM's extensive global advertising campaign with the slogan 'The other IBM', coupled with the phrase 'on-demand computing', which is simply IBM's trade name for application services based on the SOA model.

Mainframes and centralized IT structures are no longer mentioned; the new message is supporting enterprise business strategies and 'on-demand business'. The hardware and other IT products needed to implement such strategies have finally become commodity items. Understanding the customer's business has become a crucial factor for the commercial success of a supplier.

Previous ERP systems, which often combined several functions for extensive commercial applications in single monolithic blocks, must now be divided into simple, relatively small software modules for the SOA environment. These modules communicate with each other via 'service interfaces', which are a familiar feature of object-oriented technology.

In networked systems, the question of which platforms host individual modules or objects is irrelevant because EAI services provide the necessary support for communication between objects. They also provide interfaces to existing ERP systems, which are regarded as hybrid objects. This allows old and new systems to be used in complementary fashion, which in turn helps users and producers amortize their investments in existing systems over a reasonable length of time and implement smooth transitions to pure SOA structures.

Operational objectives of SOA

In a SOA environment of this sort, operational application systems communicate via standardized service interfaces instead of mutually agreed bidirectional interfaces. This makes it possible to achieve certain operational objectives:

- **Flexible system configuration.** In the SOA model, systems constantly reconfigure themselves according to the demands of the currently needed scope of services. This results from dynamic communication between application objects via SOA interfaces.

- **Scalability.** The platforms hosting these modular systems can be adapted flexibly (scaled) to the demand for services. Only as much hardware and as many system components as are actually used have to be provided. Maintaining spare capacity, which is often the case with traditional ERP systems, is no longer necessary.

- **Effective reuse.** In a SOA environment, functional blocks in the form of modules are available for constant reuse via service calls.

- **Easier maintenance.** Maintenance (software updates) and upgrades are always difficult in monolithic systems because every system modification affects the entire system and can even lead to new errors. By contrast, maintaining small service modules is much easier and less prone to errors. New functions can easily be added to the system as independent modules, and they are immediately available to all system users.

- **Flexible distribution.** Distribution across various system platforms can be performed dynamically, even including relocating decentralized services to a central computer centre, without having an impact on other areas of the system and without having to perform extensive integration tests each time. The EAI functional layer provides flexible, system-wide service links.

SOA modules do not know anything about the internal operation of other modules or the specific system environments in which they are located. They only know how to invoke each other. It does not matter whether all the modules are located in a single hardware system or linked via the network. They can also be implemented on different platforms at different times due to changes in the system configuration. A more comprehensive description of an SOA is provided by the Progress white paper [Progress2508], among other sources.

As the Internet is most often used as the network, the support components for SOA are also called Web services. Individual services and software modules can be generated according to process requirements using software development tools. The components of a SOA consist of web services, XML, message queuing, workflow support, and so on.

Supporting strategic enterprise objectives

SOA would not draw any attention if it were just a technical toy for software experts. However, SOA is expected to provide enhanced support for strategic enterprise objectives in future IT systems in the following areas:

- **Improved operational efficiency of enterprises.** Faster adaptation of information systems to changes in organizational structures and business processes, and reduced redundancy of system modules.

- **Improved ease of integration with company mergers.** Easier integration of systems built independently on different platforms using different technologies.

- **Faster information flow between business partners.** Optimization of value chains and supply chains by simplifying provision of the functions and data needed by business processes spanning enterprise boundaries.

- **Improved customer satisfaction.** Better communication with customers, who will increasingly collaborate with suppliers via the Internet.

- **Global information availability.** Supporting globally distributed organizations in optimum pursuit of their business activities.

- **Reducing software maintenance and modification expenses.** If software takes the form of small, independent functional units, it can be maintained and extended much more easily and thus more economically.

SOA is thus not a technically driven concept originating from software experts, but instead a response to the requirements of business users of software systems. As the previously described historical evolution of SOA shows, this resulted in the refinement of technical innovations such as object-oriented technology into commercially mature technologies.

Consequently, every enterprise that is dependent on IT systems must become familiar with the SOA model, even if the SOA is already included in the product.

The quality and performance features of specific SOA services are defined contractually in service level agreements (SLAs) between suppliers and users. This is also worthwhile when both parties belong to the same company. Suppliers cannot properly plan, configure or provide the system resources necessary for a particular service unless the parameters of the service are described in specific terms by mutual agreement.

The advantage of SOA is that it is an architecture that provides flexible support for system growth. With a monolithic model, increasing system performance requirements quickly lead to the need for a different computer. In an SOA environment, another server is simply installed in parallel with the existing server and linked to it via the network.

6.1.5   RFID Architecture and Event Processing

RFID reading processes occur in locations such as warehouse receiving areas fitted with RFID antennas. The antennas can recognize objects delivered to the warehouse and forward the data obtained from these objects to the warehouse management system, which is part of the central ERP system of the enterprise.

A specific system structure has evolved to ensure that this form of data provision works reliably. The portion of this structure devoted to the antenna is shown in Figure 6.5. The antenna first communicates with the reader, which is a small computer that accepts the received data and controls the transmit and receive functions of the antenna. Several antennas can be connected to a single reader. If a warehouse is fitted with RFID antenna gates at the loading doors (hubs), an edgeware server

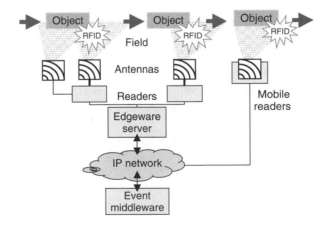

Figure 6.5   System configuration of RFID antennas for recognizing moving objects with RFID tags

is usually installed on site and all readers are connected to this server. Here 'edge' means that this is the edge of the enterprise's communication network, adjoining the real world of objects.

In a relatively small installation, the edgeware server functions can also be provided by the reader, so the edgeware server does not appear as a separate unit in the system. In the case of a mobile RFID reader, the reader and antenna are combined in a single unit. In any case, the data is fed from the edge to the enterprise network and supplied to the ERP systems operating in that network, such as a supply chain management (SCM) or warehouse management system (WMS). However, the data is first preprocessed by the event middleware, which is described below.

The reader and the edgeware server must also filter the data coming from the antennas to eliminate redundant data. For instance, objects in the warehouse receiving area may pass through more than one antenna field and thus be read more than once, or they may remain in an antenna field for a while and be read several times in the same place. From an operational perspective, what is important is simply that the object has arrived at the warehouse. All duplicate read data must be eliminated. This is best done in the reader or the edgeware server before the data is transferred to the network. In addition, the edgeware server must be able to buffer the read data until it can be transferred in complete form to the next higher system level. This also allows the RFID readers to continue working even if there is a network malfunction or the network is out of service.

Event data

RFID systems and auto-ID systems in general receive object data in real time. This has led to a new designation for such data: event data. An event is a specific situation that occurs in a specific place at a specific time. It is transferred from a reader to the IT system in the form of a message generated when the reader recognizes the event in its antenna field, which means in real time. If the antenna or reader is not active when the object moves through the antenna field, it may not be recognized, and in this case the 'warehouse entry' event information is lost. If the reader accepts the data in good order from the antenna, the edgeware server must also be prepared to accept the event message from the reader, since most readers do not have enough memory to buffer large amounts of data.

Temporary unavailability of system resources can cause incomplete data acquisition, leading to incorrect inventory records. Consequently, simply installing antennas and readers is not enough to implement an auto-ID system. An auto-ID system makes new demands on system availability, which in turn impose new requirements on the system architecture.

Central enterprise systems that support operational applications cannot satisfy real-time requirements. They are designed for users who read data

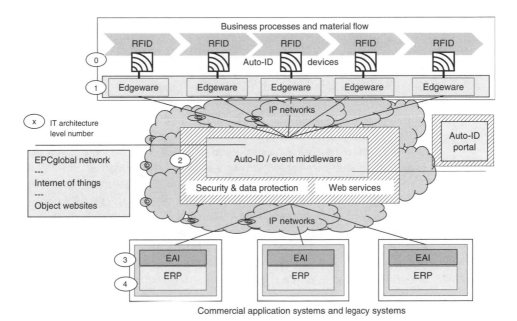

Figure 6.6 Event middleware as the middle layer of an IT architecture

from documents and enter the data via terminals and users who address queries to the system in order to retrieve stored documents, data or statistics. If terminal service is temporarily unavailable, users take breaks or work on other tasks. When the service is again available after a few minutes (or a few hours), users resume their work at the terminal.

This is not possible with events, which must be processed correctly in real time. This has led to the development of a new component of the IT architecture: event middleware (see Figure 6.6). This can best be implemented using the SOA infrastructures described in Section 6.1.4.

Other auto-ID technologies

Up to now we have been talking about RFID tags, RFID antennas and RFID readers. RFID is only one example of auto-ID technologies. Other types of auto-ID technologies operate in a similar manner in terms of data acquisition. In the rest of this book, they are indicated by the symbols shown in Figure 6.7.

Figure 6.7 Symbols representing ID technologies

Barcodes. Barcodes occupy an ambiguous position in this context. They can be recognized electronically by a scanner, but only if the scanner can see the barcode label. If the barcode is not visible, the object must be manually oriented so that it is visible. This is naturally a source of errors, for example if 11 of 12 cases are placed correctly on a belt and recognized by a barcode scanner, but the twelfth case is not recognized. In this case it will remain unrecognized, and the system will not know its actual location.

Iris scanner. An iris scanner, such as is used to scan the irises of air travellers, recognizes the pattern of a person's iris instead of an RFID tag. The iris pattern acts as an identification code. The scanner transmits the pattern to the next higher processing level of the IT system, where the identity of the person associated with the read 'iris code' is determined and a decision is taken as whether to allow the traveller to pass through passport control and board an aircraft. This works the same way as using smart cards for access control. Here each person is assigned a virtual identity that is stored in a database.

WLAN. As a medium, WLAN appears here in two different forms. The first form is naturally as a medium for connecting readers to edgeware servers or network routers, and the second form is in the role of a RFID antenna. This can be illustrated by the example of an airport that wishes to use electronic means to determine the locations of baggage carts used by passengers to transport their luggage. At first glance, fitting the carts with RFID tags appears to be a reasonable approach. However, this would require providing full RFID antenna coverage of all areas of the airport where the carts could travel, which would be a considerable expense. Most likely the airport already has a WLAN system with suitable antennas so its personnel and travellers can use their laptops at the airport. A natural solution would thus be to fit the carts with WLAN tags and use them to determine the cart locations.

GPS/GSM/GPRS. GPS devices are used when it is necessary to track the movement of relatively large objects such as containers, vehicles and semitrailers. Here there is a certain amount of overlap with systems that employ active RFID transponders, which are also used to label vehicles. However, this is only worthwhile if the control points (such as car parks and site accesses) are fitted with RFID antennas.

Comprehensive tracking can be achieved with GPS. For this purpose, the GPS device is fitted with a GSM module that communicates the geographic position determined using GPS to the tracking IT system. There the data arrives at the event middleware level, where data from other auto-ID read processes is also received. From the view of the middleware, the type of auto-ID technology used to acquire the data does not matter.

GPS methods are discussed in section 7.5 of this book. Naturally, it is also possible to locate the positions of mobile telephones via the GSM network.

6.2 Software Agents[1]

When we talk about RFID systems, the first thing we usually think of is RFID tags and objects that can be read by antennas so the data they contain can be transmitted to a higher-level system. However, there is a visible trend toward fitting objects with tags that have built-in processors with logic processing and data storage capability instead of using simple RFID tags. These processors can also make their identities known to each other via RFID. Agent systems of this sort are becoming increasingly important for software development in distributed systems. They are credited with having universal application potential. In the auto-ID environment, for example, agents are used for autonomous control of load carriers in industrial conveyor systems. How this works is described here in detail.

This section provides an introduction to the technology of software agents and presents the underlying terms and concepts. It also compares agent systems with alternative methods. In this regard, particular emphasis is given to the architecture in which the agent system is embedded, which represents an extension of the overall architecture of an operational information system.

6.2.1 Terms and Definitions

The term 'agent' (from the Latin for 'doer') comes from the field of artificial intelligence (AI) research. It describes a closed technical system that interacts with its surroundings and independently makes decisions in the process.

Software agents, which are henceforth referred to as agents, are independently executing software programs with predefined, application-specific goals and interfaces for exchanging data or messages. In a multiagent system (MAS), which is an association of several agents, agents exchange data with each other and thus perform cooperative activities in their environment.

The actions of an agent alter its environment and thereby influence its future decisions. Of course, a prerequisite for this interaction is that the agent has suitable information about its environment. Performing actions

---

[1] Contributed by Dirk Liekenbrock of the Fraunhofer Institute for Material Flow and Logistics (IML), Dortmund, Germany (Director: Professor Michael ten Hompel). http://www.iml.fraunhofer.de

and communicating with other agents are not possible in an unknown environment[2] [Fer01].

Software agents are characterized by several properties that are presented here in the form of a definition. According to this definition, an agent is a software program with one or more of the following properties:

- **Autonomy:** working independently and largely without user intervention.
- **Proactivity:** triggering actions on its own initiative.
- **Reactivity:** responding to changes in its environment.
- **Social behaviour:** communicating with other agents.
- **Ability to learn:** learning from previous decisions and observations.
- **Adaptability:** adapting to changed environmental conditions.

In the original object-oriented technology, agents would have been called 'objects'.

To support interaction with its environment, an agent is equipped with a sensory system for receiving external information (such as ambient temperature) and a knowledge base (database). Intelligent agents exhibit all of the above-mentioned properties. They are distinguished by their knowledge, learning capacity, ability to draw conclusions, and ability to alter their behaviour. Further subclassifications can also be identified.

6.2.2   Types of Agents

Not all agents have all of the above-mentioned properties in fully developed form. Among the many different types, two can be selected that differ in the knowledge they manage.

Reactive agents

These agents do not have an extensive knowledge base, but instead act directly on the basis of information from their environment. Two types can be distinguished:

- A **reactive agent** is the simplest type. It uses condition/action rules to select an action based on available sensory information.
- A **teachable agent** extends the capabilities of a reactive agent by having a memory that stores the effects on its environment resulting

---

[2] Environment information can also include the circumstances of the intended execution platform (hardware and services) of the agent.

from its previous actions. The condition/action rules interact with the memory information and the sensory information.

Cognitive agents

A cognitive agent maintains a model of its environment in its internal data structure. This enables the agent to plan its actions and act in a goal-oriented manner. A well-known representative of this type of agent is the BDI agent architecture.[3] The following types can be distinguished here:

- A **goal-oriented agent** has a predefined goal. Based on sensory information and the predicted consequences of its actions, it selects an action that leads towards its goal.

- A **utility-based agent** is a further refinement of a goal-oriented agent. It also has a predefined goal. In addition, it can predict consequences over multiple decision stages and generate plans in this manner. This is especially interesting in situations where it is uncertain whether the goal can be attained. The agent performs weighted risk assessments and does not always pursue only its primary goal.

In contrast to programs that are immovably installed on individual computers in a network and are thus static agents, the underlying idea of a MAS implementation is mobility of the agents, which means transferring executable program code via a network.

Mobile agents

A mobile agent is an executable process that independently migrates among the computers of a network of heterogeneous computers. The agent uses the services of its current host computer and can communicate with other local agents and agents on other computers. Agents can be instantiated multiply. The multiple instances of the agent are launched as independent processes, and they perform a task concurrently and *cooperatively* on different computers.

Agent platform

Mobile agents can move from one computer to another in a network. An agent platform installed on the participating computers provides the agent with an execution environment abstracted from the operating system and network, which enables the agent to access resources (memory and CPU) and transport services (communication). The platform also provides other services, such as login/logout, thread control and access control.

---

[3] Beliefs, desires and intentions.

Agent system

An agent system consists of several agent platforms distributed over a heterogeneous network. Ideally, but not necessarily, the same system platform is installed on all of the computers in order to ensure portability without any further requirements (see Section 6.2.3).

6.2.3   Agent Concepts

Agent systems typically have properties such as mobility, migratability and/or portability.

Mobility

In order for an agent to be able to switch platforms, it must be possible for the executable code to be transferred between the platforms. Two types of migration can be distinguished:

- **Code on demand.** The program code and data are transferred from a server to a client, which then executes the code. Java applets are an example of this.

- **Remote execution.** In this case the program code and data are transferred from a client to the server, which then executes the code. Java servlets are an example of this.

Both cases are characterized by the fact that the transfer of the executable code is triggered by the system. By contrast, the transfer can also be initiated directly by this agent. In this case, the agent can migrate.

Migration

In the case of migration, agent execution on the current hardware is suspended and the agent is transferred to the target platform and then deleted from the previous platform. Agent execution is resumed on the target platform. Here the agent manages three components: the program code, the data state, and the execution state.

In the case of **strong migration**, all three components are transferred to the target platform. After the execution state has been determined, the components are converted into a transport format (serialization) and transferred. In the case of a heterogeneous target platform with different hardware, this can only be done if an independent representation of the execution state is possible. Strong migration thus represents a major challenge.

Weak migration differs from strong migration in that only the program code is transferred. After the transfer, the agent is launched on the target platform and its execution state is determined from its data state. This agent architecture corresponds to a state machine that is initialized with its most recent execution state when it is instanced.

Here a distinction must be made between the internal and external states of the agent. The internal state encompasses code execution (program counter, stack and memory), while the external state encompasses access to the input/output (I/O) units of the target platform, for example for file access or communication. The external state is usually a critical issue with migration between heterogeneous platforms. Consequently, most agent systems only allow transfer of the internal state.

Portability

Migration in heterogeneous systems with different types of platforms is only possible if the program code of the agent is portable. This requires a virtual machine and an interpreter. This arrangement abstracts the execution environment of the hardware, and the agent code is executed in a restricted environment.

Security

An agent is a foreign program that must run on the target platform within the framework of specific security guidelines. These guidelines determine which operations can be executed and define the authentication process. There are three aspects of security:

- **Protection against unauthorized access.** Agent access to the resources of the host computer must be controllable in order to prevent the agent from performing potentially harmful operations. This requirement is typically fulfilled by using a virtual machine (VM), which also governs resource utilization.

- **Data security.** Encryption mechanisms can be used to protect an agent system against counterfeiting and spying out the code or data transferred during a migration.

- **Interoperability.** If several agents run concurrently on a single platform, they must run without interference. Consequently, the agent platform must monitor memory use and memory access to prevent resource conflicts.

Communication

Three types of communication can be distinguished in agent systems:

- The **agent** and the **platform** must exchange information with each other so the agent on the agent platform knows which services are available and what resources the platform has. This includes handling exceptions and errors.

- Message exchange **between two or more agents** is used to coordinate cooperating agents in the network. This can occur locally on a single platform or in the system via the network.

- Communication **between platforms** is necessary for agent mobility. It is intended to ensure fault-free migration and localization of agents.

Robustness

Failure of individual computers, network failure and data transmission errors can occur during normal operation of an agent system. If possible, a MAS should be robust with respect to such faults, and execution of the agent application should not be affected by them. For this purpose, an agent system must include mechanisms that continue execution of the affected agents after a failure or malfunction. Furthermore, agents must not disappear or be duplicated during migration. Migration security can be supported by transaction mechanisms.

6.2.4   Applications, Architectures and Benefits

The concept of agent systems is only partially new with regard to the hardware and software components of such systems. As with all software programs, agent systems consist of a combination of algorithms and data structures. An agent system acts to process user data and perform computational processes.

What is new is strong dispersion and task orientation. The architecture of an agent system is designed such that individual agents take on clearly demarcated tasks, either alone or in association. They thus represent an extension of the object-oriented software concept with the addition of the elements of dispersion and intelligent behaviour.

Applications

Potential applications for agent systems can predominantly be seen in handling complex tasks that cannot necessarily be handled by centralized IT architectures. This includes planning and control tasks in which a large number of individual activities must be coordinated, as well as automation of business processes related to the strategic objectives of a real-time enterprise.

One example is planning production and logistics processes, with orders and resources being represented by agents that independently manage order progress. This includes allocating resources to orders and scheduling multiple orders. The planning steps are executed using a negotiation scheme, such as an auction that lets agents bid for the use of specific resources. Instead of a central planning entity, which would have to process relatively large amounts of data given the number of orders, planning is decentralized in an agent system.

An advantage of agent-based planning or control is a significant reduction in the complexity of software development. In contrast to centralized architectures, in which business process control is implemented in a single location (the server), an agent system allows this control to be distributed and divided into easily understandable functions.

Decentralized processing of order data is also an element of client/server systems. In such systems, a client sends an order with the associated data to a server via the network, and the server executes the order (Figure 6.8). After data processing is completed, the results are returned to the client. All client requests are processed centrally by the server, which in this star topology is thus a critical component with respect to availability and which must handle requests from multiple clients concurrently.

The client and the server exchange messages synchronously via the network, so they must constantly maintain links to the network. It is not uncommon for large amounts of data to be transferred from the server, even though only part of the data is used by the client (e.g. with database queries). This causes the full capacity of the network and server to be used exclusively by one client for a certain time.

Automation hierarchies

Centrally organized architectures prevail in the domain of automated machinery and systems. In such systems, the role of the server is assumed by a control system or production control system, which controls the process in the automated system (Figure 6.9). The components for

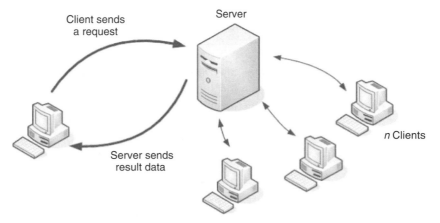

Figure 6.8 Transactions in a client/server system with a star topology (source: Fraunhofer IML)

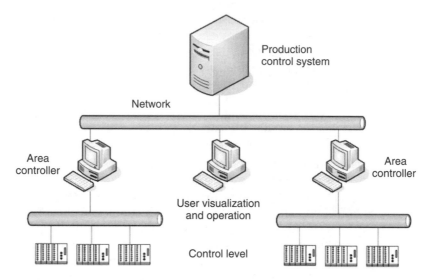

Figure 6.9   Level structure of production automation systems (source: Fraunhofer IML)

machine-level control, area control, display, user command and data entry, and the master control system are networked over several levels.

The requirements on the components and implemented software must be clearly demarcated at the individual levels. For instance, the control level is characterized by small data processing volumes and the ability to handle real-time events with a guaranteed response time, while the area control system and the master control system process larger volumes of data in the nonreal-time domain.

Benefits

Agent systems can be used to implement cooperative solutions to the task requirements of the mapped systems in distributed environments. Extensive communication with a central node (server) can be minimized by having the client process generate an agent and transfer it to the server. There the agent can perform its task in a prepared execution environment and communicate locally with suitable services and other agents on the server. Finally, the agent can return to the client with its results.

In an agent system, processing methods are transferred via the network instead of raw data. These methods are used for decentralized processing of the data of the network participants. This can sometimes lead to dramatic reductions in network loading. Agent systems thus enable simpler, faster and more economical development of distributed systems. The asynchronous communication is especially suitable for networks with small bandwidths, and it does not require fixed connections. This means that agent technology also supports mobile applications without constant network connections.

Pervasive use of agent solutions must take processing speed requirements into account. In particular, real-time capability is especially important for control tasks at or close to the process level.

6.2.5   Standards, Frameworks and Platforms

Standards

The previously mentioned CORBA, which is a platform for object-oriented software systems, plays a substantial role in the standardization of agent systems. CORBA is a language-independent standard for formal specification of the classes, objects and all parameters and data types of distributed systems. Uniform communication between the objects of a software system is provided by an interface definition language (IDL). The preferred programming languages for CORBA are Java and C++, but there are also implementations for many other languages. An IDL compiler generates a source code skeleton for applications with uniform identifiers but without any application-specific program logic. CORBA provides a set of services for locating objects, invoking remote methods (remote procedure call, RPC), migrating objects, and several other things. A design language called Unified Modelling Language (UML) is commonly used to develop objects, while XML is used to describe the documents used to exchange data between agents.

Frameworks

Frameworks are development environments for MASs. There are presently more than 100 frameworks that have been published and are used commercially. Consequently, we can only provide a brief description of a few well-known frameworks here.

Agent Builder (http://www.agentbuilder.com) is a commercial framework for developing MASs with Java. It supports various communication mechanisms, including Knowledge Query and Manipulation Language (KQML).

The **D'Agents** framework essentially consists of two parts: a server for mobile agents and a modified Tel interpreter. Tel is a tool command language for rapid application generation. This research is funded by military organizations, including the US Department of Defense (DoD).

Java Agent Development Framework (JADE) (http://jade.cselt.it) is coded entirely in Java. It simplifies MAS development by providing middleware and a set of graphic debugging and development tools. The middleware conforms to the FIPA specification. The agent platform can be distributed over machines with different operating systems.

Secure Mobile Agents (SeMoA) (http://www.semoa.org) is a framework with the emphasis on security. This open-source project provides security mechanisms for mobile agents.

The **Aglets Software Development Kit (ASDK)** from IBM Tokyo Research Labs is an environment for programming mobile agents in Java. Aglets are software objects that can migrate between hosts in a network. ASDK is available free of charge (see http://sourceforge.net/projects/aglets), but it is no longer being developed.

ZEUS (http://sourceforge.net/projects/zeusagent) is an open-source project developed as part of the Agentcities research project, primarily in cooperation with British Telecom. The agent architecture is based on the BDI model.

Platforms

An agent platform is a physical infrastructure that provides an execution environment for agents. A platform on a computer system provides functions for agent execution. This includes an agent management system (AMS), a directory of agents (Directory Facilitator, DF) and a communication channel (Agent Communication Channel, ACC). Standardization of agent platform implementation is only possible to a limited extent due to the application-specific focus of agents.

The aforementioned Java-based frameworks support implementation of MASs on a virtual machine provided by a Java runtime environment.

6.2.6   Future Prospects

From identification technology to autonomous objects

Knowledge of inventory levels, goods locations and the quality of goods is a basic requirement for fashioning efficient logistics processes. Continual identifiability of goods in material flows is a key function. RFID systems have already achieved a level of technical development that enables implementation of not only the basic identification function but also relocation of material flow control functions.

On the way to real-world awareness (RWA)

RFID tags are already being used to store the process data of individual logistics objects on their RFID tags. The benefits of RFID in material flow control can presently be seen especially well with returnable container systems in which the tag information is used to control container travel past the intermediate stations of internal transport, restocking, picking, and provision for shipment.

The following description is based on experience with the laboratory setup of the University of Dortmund, which models a continuous conveyor system (Figure 6.10). Conveyor systems of this type are used in distribution systems and baggage conveyors at airports to provide automated container transportation. Controlling such systems is a very

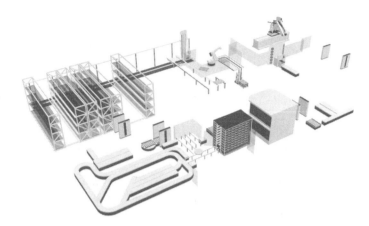

Figure 6.10 Laboratory setup for intralogistics applications (source: Fraunhofer IML)

complex task because it requires coordinating many individual parameters, such as routing, empty container management, and load balancing during peak periods. RFID tags can store part of the necessary information and thus allow decentralized control structures to be used. Initial experience in the industrial environment is already available, and it shows that decentralized architectures are feasible.

Allocation of tasks to the operational units of the logistics chain is decentralized in such architectures. This can be illustrated by means of the following description of a typical process chain.

After identification of the containers delivered to the goods receiving area on a pallet, the containers are first transported to the returnable container depot. The status of each container is determined by manual inspection, and an RFID reader is used to write the status to the container's RFID tag. After this a conveyor system sorts individual containers onto type-specific belts. This sorting is performed fully automatically using the container information in the RFID tags.

After the containers have been loaded on pallets, a request to send a driverless transport vehicle (DTV) to the pallet location is sent to a transport control system. After picking up the load, the DTV reads the storage area information previously written to an RFID tag on the base of the pallet and navigates to the warehouse transfer station. There the pallet with the containers is identified by an RFID antenna portal, after which the inventory management function of the associated warehouse management system determines the coordinates of a free storage location and writes them to the pallet tag. The DTV then deposits the pallet at the designated location.

Additional operational units can be chained in the same manner, and this chaining illustrates how process data attached directly to goods can be used for control tasks. Because of decentralized data management, the number of transactions in the central inventory management system is reduced significantly. Although all inventory changes are also maintained in the central system, it is no longer necessary to request object and process information from every identification point for this purpose.

Without RFID, item identification is performed by reading ID codes (such as barcodes) and linking item data to process data by retrieving the data from a central database. RFID paves the way for a transition to decentralized, near-real-time data management and data processing using RFID tags on logistics objects. This is often called RWA. RWA is a designation for automatic linking of real physical objects to control software.

Besides RFID technology, RWA requires a suitably adapted software architecture. This typically involves associating a software agent with each entity of a material flow system, which includes physical objects as well as resources. Each software agent can be instanced on any desired hardware platform, and it only needs a data link to a reader. The agents independently organize the subsequent process steps in the sequence of events between two identification events for the associated objects.

Software agents and RFID: aware objects

With each identification transaction, the software agent receives the current location and status information of the associated object based on the data in the RFID tag, calculates the future process data in cooperation with the other agents in the system, issues an instruction for the next process step, and initiates writing of the data to the object's tag.

Fixed software agents

Agent systems create a virtual model of the entities of a material flow system and control the course of real physical processes via interfaces. As each agent operates independently and has its own goals, the result is a highly distributed system. From a technical perspective, this system architecture is interesting because it provides an opportunity to model a wide variety of principles of self-organization of autonomous entities. The range of variation of the resulting implementations is broad and extends from relatively simple dispatching methods to cooperative negotiation models with blackboards, and even to very elegant but behaviourally complex methods analogous to natural phenomena.[4] Known methods of discrete mathematical optimization are also used here.

RFID tags with sufficient memory capacity to hold the necessary data and control system models for systems using software agents are a characteristic technical requirement for self-organizing systems. Ideally, each agent−object pair should form a physical unit as well as a logical unit.

---

[4] Proven methods include modelling of swarms. Some examples taken from nature are ant swarm and bee swarm algorithms, which in technical terms implement distributed routing in networks [Gam99].

Mobile software agents

Transferring the concept of software agents to RFID tags and subsequent creation of objects that can operate autonomously requires programmable tags with internal energy sources to ensure that they can operate outside the antenna fields of readers.

This results in tags with extended functions. For example, environmental parameters such as temperature can be measured by sensors connected to the tag, and the measured values can be processed. If the maximum allowable temperature of a food is exceeded during transport in a cold chain, this can be recorded and processed locally. When the tag of the load unit is within range of a stationary reader, this information can be transferred and an alarm message can be generated.

Active tags also allow data to be exchanged with other tags in their surroundings. Such 'aware objects' [tenHo05] can communicate directly with each other to request transportation resources or negotiate processing sequences (see Figure 6.11) without any need for an intervening infrastructure.

Wireless sensor networks

Full self-organization is achieved when the identity, current status and position of every individual object (container, pallet, product, etc.) are known at all times. Complete decentralization of material flows requires several technologies in equal measure of importance: identification technologies, sensor technologies, communication technologies, and security technologies.

Wireless sensor networks are a potentially suitable platform for implementing aware objects [Kok04; Mül05]. Suitable highly integrated devices

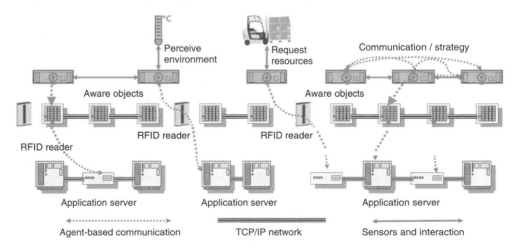

Figure 6.11   Interaction and communication in the internet of things (source: Fraunhofer IML)

with small sizes are already available, including devices with dimensions of a few centimetres containing highly integrated electronic structures and built-in antennas. If they are provided with internal energy sources or supplied with energy via electromagnetic fields, such devices can operate autonomously for extended intervals. In addition, research efforts are already developing techniques for polymer tags, which can be produced in successive layers using printing processes and attached in the form of film labels.

There are already devices available with a programmable processor, working memory, and interfaces for connecting additional sensors. Devices can establish links to neighbouring devices via a wireless network and exchange data with each other. Position information can be calculated using localization methods that determine the distances between at least four nodes by measuring propagation times and calculating relative coordinates in the associated network. In order to determine absolute spatial coordinates, it is also necessary to have stationary nodes with fixed, stored spatial coordinates to serve as reference points.

As ad hoc networks, wireless sensor networks can dynamically integrate new nodes into the system and establish the most economical communication link in each case in order to minimize the energy consumption of the nodes and achieve longer operating times.

Future prospects in the internet of things

From a technical perspective, development of RFID technology on the way to the internet of things is already feasible. The necessary components are available, and demonstration environments that can be used to study and illustrate what can be done with these components are the subject of current research projects. Aside from hardware and energy management issues, a critical area of research for initial implementation of the internet of things is in suitable algorithms for localization, autonomous control and self-organization. The internet of things need not always be a global structure, as it is usually presented in discussions. It can also exist on a small scale inside an enterprise, especially when the real world of processes and the virtual world of IT systems can communicate with each other in real time via auto-ID processes without manual intervention.

Increasing integration density

This evolution is accompanied by research aimed at achieving higher component integration densities and more economical manufacturing processes. Here we can highlight activities aimed at creating polymer-based RFID tags that could be applied to substrates using printing methods and would thus have very low production costs. These 'tags on a roll' could be produced with unit costs in the cent range, which would create the economic conditions necessary for widespread use of RFID technology. Research is also being conducted on miniaturization of silicon-based IC technology and fabrication of tags (chip plus antenna) and associated interconnection elements for use in 'e-grain' applications.

Higher connectivity

The described evolutionary steps form the technological basis for achieving higher connectivity of logistics objects. The goals of improved traceability and visibility in the sense of near-real-time data processing in logistics systems can only be realized if continual access to the data of individual objects is possible. This is exactly where the analogy to the Internet becomes clear, since the same mechanisms are necessary for locating, routing and tracing physical objects in the internet of things. This always implies direct communication by physical objects, which represents the concept of RWA [Hein2003; Hein2005].

6.3   Real-Time Enterprise Infrastructure

6.3.1   IT Architecture

Use of auto-ID methods for automatic transfer of real-world data to information systems is an essential characteristic of a real-time enterprise (RTE), which has continual access to all data in its information systems that are needed for controlling processes or indicating when processes are not running properly and require near-real-time control actions. If this data is not available in real time, as with non-RFID methods, it is still possible to analyse the causes of errors but too late for control actions.

The IT architecture of an RTE requires a significantly larger middleware component. Its function is no longer limited to application integration (EAI) in the style of ERP systems, but instead includes preprocessing of event data arriving from the auto-ID periphery, which is why it is also called 'event middleware'. This is where event data is processed to obtain operationally significant data that can be passed on to central ERP systems. An event is a message that arrives at an auto-ID reader when its antenna recognizes an object with an electronic identity, such as an RFID tag. ERP systems cannot handle such events directly. A comprehensive general architecture suitable for supporting an RTE is described below. It consists of five layers (layers 0 to 4) as shown in Figure 6.12.

This architecture includes not only RFID technology, but also other auto-ID technologies such as barcodes, GPS/GSM, WLAN and the like, since they are all suitable for acquiring data in real time. Other technologies such as iris scanning and face recognition can also be used for this purpose, but they are not considered here. The symbols for the various types of auto-ID technology are shown in level 0. The edgeware

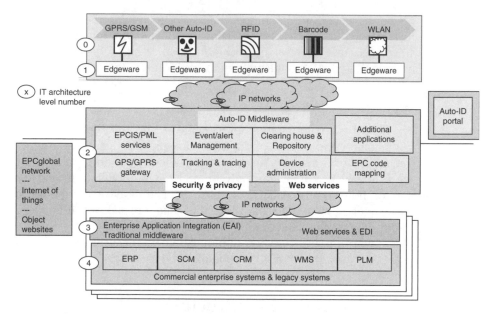

Figure 6.12　Functional levels and components of an RTE information architecture

servers are located in level 1. They forward events via the Internet to the auto-ID/event middleware in level 2. The technology used to generate the data does not matter to the event middleware. All it cares about is the associated events and when and where they occurred.

Previously, middleware (or EAI) was only used to link application systems to enterprise servers or central mainframes (level 4) in the enterprise or to global communication networks. Auto-ID middleware provides completely new functionality for event processing. The previous middleware is thus divided here into an EAI level (level 3) and an event processing level (level 2). The components of the IT architecture shown in Figure 6.12 are described in more detail below.

Level 0: auto-ID peripherals

Readers and antennas that recognize the electronic identities of objects. The underlying technology can be RFID, WLAN (WiFi), barcodes, etc. The data is passed on to the edgeware in level 1.

GPS devices on mobile objects such as containers are used to recognize geographic positions and pass this information to the middleware via a GSM/GPRS gateway.

Mobile units based on GPS or RFID tags can also be configured to receive, store and convey sensor data. For example, an electronic lock on a container, the temperature history of a cold storage container, and the operating parameters of the vehicle can be monitored. This data can also be transmitted via the auto-ID link.

Level 1: edgeware

This level contains the edgeware servers, which are located close to the auto-ID peripheral equipment and connected to it by cables or WLAN. An edgeware server monitors several readers or antennas. The term 'edge' reflects the fact that the data is fed into the extended auto-ID network here. The tasks of the edgeware servers are:

- monitoring the operational status of readers and antennas;

- generating alerts in case of operational problems;

- initial validation of auto-ID data (events) and filtering multiple data sets resulting from multiple reading of objects that pass through several antenna fields;

- forwarding operationally significant events, such as 'Pallet W has entered portal X' or 'Palette W holds Y cases of product Z';

- buffering data if the next-higher system is temporarily unavailable for communication.

There are also reader configurations without individual edgeware components, such as mobile readers and RFID-enabled mobile telephones (NFC mobile phones). The antenna is also integrated in the housing with such devices. If the edgeware functions are thus not available, they must be provided at the middleware level during event processing. In this sense, the division of functions between edgeware and event middleware is fluid.

It must be possible for the event middleware to verify the functionality and operational status of the distributed auto-ID peripheral devices. This ability is provided by the device management component at level 2.

Level 2: event middleware

This is a new application layer for auto-ID. It belongs to the middleware in the traditional classification of system levels, but it represents a new, independent application layer. The following functions must be implemented here:

1. **Device administration.** Registration of auto-ID devices at the periphery. System communication with the edgeware or readers in order to recognize their operational status and verify their correct operation. Generating alerts when devices do not operate properly or links to devices are interrupted.

2. **Event and alert management.** Receiving, storing and processing event messages. Preparing messages for forwarding to ERP systems,

checking events for exceptions and generating alerts. This is the actual processing layer and the location of functions such as aggregation and disaggregation of load carriers (such as containers and pallets).

3. **Tracking and tracing.** Storing the planned and recognized routes of objects. Answering search queries for such objects. Extraordinary events (such as deviation from a planned logistics route or unauthorized movement of an object) are also recognized here. If such an event is recognized, an alert is sent to previously defined locations.

4. **EPC code mapping.** Maintaining EPC tables and converting codes between sector standards as necessary. This function is useful if the system monitors system processes that span sector boundaries, as it can be assumed in such cases that different code standards will be used in different sectors.

5. **EPC information services (EPCIS).** Access to the EPCglobal network. A special dialect of XML called Programmable Markup Language (PML) is used for this purpose. EPCglobal also uses the term 'EPC Discovery Service' (EPCDS) in this connection.

6. **Clearing house and repository.** Clearing means filtering data according to intended message recipients at the ERP level. Data storage, database administration and ensuring that data queries are answered properly. Clearing payment transactions for settlement of provided goods and services. Based on a global repository (database) containing all object, route and participant data necessary for the supported processes.

7. **Value-added applications.** These are primarily statistics applications for use with the database accumulated in the repository.

8. **Auto-ID portal.** Provides access to the system for authorized users, such as:

   - administrators;

   - logistics users who wish to track or trace an object;

   - object users (including consumers) who wish to access an object website in order to obtain product characteristics, maintenance information or best-before dates;

   - invoking value-added applications, such as statistics for evaluating logistic routes.

9. **GSM/GPRS gateway.** Monitors message traffic generated by mobile devices via GSM/GPRS links, such as GPS devices, mobile RFID readers and RFID-enabled mobile telephones. Monitoring this traffic forms part of network management.

Level 2 is implemented entirely in the Internet (IP network). Web service components are used for communication control. System security is achieved by using privacy and security methods.

Level 3: EAI middleware

- **Enterprise application integration (EAI).** This is the conventional middleware level, which is used for application integration inside an enterprise, via the network, or with trading partners. It is physically located close to the ERP systems, such as in computer centres. It is implemented in the Internet by Web services software.

- **EDI.** An EDI component is shown at this level. Electronic packing slips, such as ASN (advanced shipping notice) and DESADV (despatch advice), are examples of EDI messages. When a load (such as a pallet) arrives at its destination, the pallet content information obtained using auto-ID can be compared with the electronic packing slip to determine whether the delivery is correct or there is a shortfall. The EDI service can also be provided by a module at the ERP level.

Level 4: ERP systems

These are operational application systems that receive operational data generated from events. This requires compliance with corresponding interfaces, such as the SAP Exchange Infrastructure (XI).

ERP systems cannot receive event data directly because they are not designed to handle the associated real-time data traffic or data volume. ERP systems include numerous components that support operational tasks, which are designated using three-letter abbreviations:

- **ERP:** enterprise resource management (higher-level classification term for this software).

- **SCM:** supply chain management (supply system).

- **CRM:** customer relationship management (marketing and sales system).

- **WMS:** warehouse management system.

- **PLM:** product life cycle management.

- **LES:** logistic execution system.

6.3.2 IT Infrastructure for On-Demand Services

There is a growing trend for enterprises that do not wish to deal directly with the complexities of IT systems to procure these systems and services from external providers or transfer operation of their computer centres to external service providers. In the past, this was usually called 'out-sourcing', but the term 'on demand' has become increasingly common in this connection because it also covers billing methods. These billing methods are based on the extent of the services actually provided. In simplified marketing language, this is also called 'on-tap IT' by analogy to beverages. The term 'on demand' was initially promoted by IBM, but even SAP, whose core business is selling comprehensive software packages, has started to offer on-demand services in cooperation with IBM. As auto-ID processes in logistics chains have an impact on several enterprises, they are an especially attractive potential application area for on-demand services.

To enable implementation of auto-ID processes, the enterprises con-cerned must establish and operate comprehensive network structures. Here there is a close connection between supplying physical objects and exchanging messages, as illustrated in Figure 6.13.

Of course, enterprises first conclude contracts that regulate the supply relationship, the products to be supplied, the terms and conditions of payment, and so on. In a modern enterprise, EDI messages are used to place orders or call up lots. EDI is a general designation for various standards used for this purpose. They specify the structures of the messages that are exchanged. They must be defined by mutual agreement so the messages transmitted by the ordering enterprise can be understood and interpreted by the receiving IT system. Commonly used EDI standards include the GS1 EANCOM standard in the trade sector and Edifact in the automotive sector.

After an order has been sent, the goods are prepared and provided to the shipping warehouse, such as on a pallet with cases containing the ordered products. When the pallet has been loaded onto a lorry, a despatch advice message is sent to the customer. The despatch advice message essentially contains the packing slip information and the departure date. If RFID is

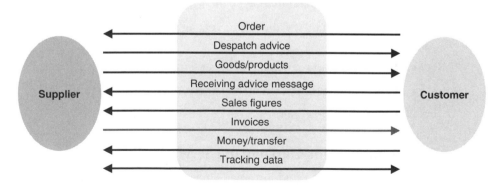

Figure 6.13   Goods and information flows in a supply chain

used, the despatch advice message is triggered automatically when the RFID reader at the warehouse exit reports that a pallet with a particular shipment number has been recognized and relays this message this to the IT system.

When the pallet arrives at its destination, the customer sends a receiving advice message to the supplier. The response to this is to issue an electronic invoice, which is in turn followed by a payment. In many cases, the supplier initiates payment directly. Information of this sort is exchanged not only between pairs of enterprises, but also with many other enterprises in cross-enterprise supply chains. This results in a communication network with a complicated many-to-many structure as shown in Figure 6.14.

This many-to-many structure includes an additional dimension, and thus considerable additional complexity, due to the necessary involvement of a logistics service provider that provides transport services. The logistics service provider in turn engages freight forwarders for transport over individual sections of the route. Shipments also pass through intermediate storage areas where individual containers or items are separated out and assembled into new loads for further transport. These shipments are also accompanied by EDI messages that advise which loads must be picked up where and when, as well as when and where they must be delivered. EDI messages are used to regulate just-in-time (JIT) production and delivery links that have been in common use for many years decades already. Ideally, each delivery arrives at its destination exactly when it is needed for production (such as fitting in a vehicle) or selling (such as shelf stocking). Disturbances to EDI traffic lead to disturbances in operational business processes. This must be avoided, and reliable operation of EDI networks is a fundamental factor in this regard.

EDI messages are necessary for using auto-ID methods. For example, receipt of a pallet at a warehouse can only be checked properly if it has been announced in advance by a despatch advice message. The

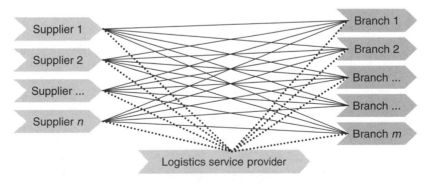

Figure 6.14 A many-to-many network structure connecting suppliers and trading enterprises with additional links to a logistics service provider

content of the pallet or container, as appropriate, is then checked against the despatch advice. If there are deviations from the correct status, a complaint can be registered with the supplier.

In order to monitor supply routes using RFID, suitable RFID antennas must be installed at locations where passage of the goods must be registered. The messages generated by the antennas must be conveyed to the supply chain management systems of the enterprises concerned so they can monitor correct progress of the shipment and intervene in case of deviations.

In the case of a shipment (such as a container) sent from the Far East to Europe or the USA, possibly with an intermediate stop in Dubai, RFID is not an adequate solution because it is not possible for suitable RFID antennas to be present everywhere. It would thus make more sense to fit the container with a GPS device that can transmit GSM messages containing the current position of the container in the form of geographic coordinates.

It is conceivable that such position determination devices on containers could be used in the future to transmit not only position data obtained from GPS receivers, but also other parameter data obtained from suitable sensors, such as temperature deviations of a temperature-controlled container or unauthorized manipulation of an electronic seal.

Establishing such networks is not difficult for wealthy enterprises such as Wal-Mart and the Metro Group or their suppliers such as Nestlé, Kraft Foods and Procter & Gamble. But what happens if widespread use of this method becomes necessary, such as if small and medium-sized enterprises (SMEs) wish to use this technology or must use it in order to comply with requirements mandated by large trading organizations? To do so, they would have to establish individual communication links with their customers and suppliers, which might involve many different interface requirements.

It appears unlikely that such enterprises will be able to provide the financing necessary to set up such complex IT infrastructures or connect their IT systems to them. Consequently, service providers will assume responsibility for this task and offer suppliers and traders simple interfaces that allow communication activities to be performed at an acceptable cost. Such services are called 'on-demand services' or 'managed services' in the contemporary language of service providers. If an enterprise uses such on-demand services, it only has to establish one communication link with a single interface. The service provider assumes responsibility for collection and distribution of messages and data processing for logistics requirements. This distinctly simplifies the network, as can be seen from Figure 6.15.

Major service providers are already presenting their first on-demand service products. It is interesting to note that some national telephone companies are also positioning themselves as suppliers, such as in

Figure 6.15 Reducing the complexity of a many-to-many network: each participating enterprise has only one communication link to an on-demand service (managed service)

England, the Netherlands and Switzerland. However, it is still unclear what functionality users can expect. This is related to the fact that there is not yet sufficient market demand for such services because the business models for using RFID are still unclear and the use of EDI and auto-ID methods is not yet sufficiently widespread. Everything is still in the preliminary stage.

Managed services in the aviation sector

A new type of auto-ID managed service has been introduced recently by SITA, the world's leading provider of IT business solutions and communication services to the air transport industry (ATI), based in Brussels, Belgium (www.SITA.aero). The strategic goals and functional contents of this solution are described by SITA as follows [Boen2007]:

> The air transport industry includes but is not limited to airlines, airports, global distribution systems, aerospace, ground handling, catering, fuelling and maintenance companies. In this industry many of the players are at the same time colleagues, competitors and partners of each other. Most ATI players operate in multiple countries and the majors do this globally.
>
> The complexity of their logistic chains increases dramatically as each airline flies to multiple destinations and their business partners are not necessarily the same in each of these locations. To add to the complexity there are more than a 1000 carriers worldwide, many of whom carry passengers and freight on behalf of other carriers and logistics providers.
>
> Due to this inherent complexity it is virtually impossible to create a transparent, end-to-end supply/logistics chain when thinking along the classical hierarchical lines: with every major company setting up their own system and getting smaller players to align (often many times over). Each player involved would end up building multiple bespoke interfaces just to interact with each other and do business. This quickly becomes expensive and unmanageable.

In addition, ATI companies want to interact with each other as peers, but while maintaining full control over their respective data. As a result, the only option allowing the ATI to adopt Auto ID/RFID technologies in a cost effective manner, is through a 'Managed Service'. These services need to support the tracking and tracing of passenger baggage as well as containers, pallets and all kinds of goods in the logistics area. Furthermore, they enable aircraft manufacturers and MRO (Materials Repair and Overhaul) companies to trace parts, control their life cycle and prevent the usage of counterfeited parts (bogus parts).

In such a managed service solution, the service provider plays the role of mediator. It does not participate in the logistic process but provides at least the following:

- A multitude of ways to connect customers' back office systems/hosts to the service.

- A simple yet powerful way of converting documents from the various parties so they can be used in the service.

- A fully configurable way in which data is shared between the various participants in a logistics chain. We call this a Value Network (VN).

- A powerful but easily configurable set of complex business rules that process signals and documents so 'right time' information is available and decisions can be made.

- A flexible way for partners to join/leave the VN.

- Integration with industry standards like ONS, DS and EPCIS.

- Ability to interact with other managed service providers.

Up to now, auto-ID methods have been used primarily in closed processes, such as production control, that are under the direct control of individual enterprises and do not involve any other enterprises. However, there are many driving forces that demand the use of such systems, including across enterprise boundaries, such as:

- further extension of requirements mandated by retail organizations such as Wal-Mart and Metro Group;[5]

- homeland security provisions (not only in the USA);

- requirements for fully traceable product life histories (e-pedigrees) in the pharmaceuticals sector and the aviation, automotive and CPG sectors.

Another important requirement on such networks is absolutely reliable handling of the data of the customers integrated in the system. Messages

---

[5] The potential cross-enterprise benefits of RFID use in the retail sector are described in [Hansen06a] and [Hansen06b], among others.

REAL-TIME ENTERPRISE INFRASTRUCTURE

transported in such networks must be differentiated such that each recipient receives only its own data and not the data of its competitors. This is not a trivial task when one considers typical supply chain situations. For instance, cases from different suppliers and for different recipients can be located in a single container. If position messages are sent to these parties, they must not find out which other shipments are present in the container. The IT systems of relatively small enterprises would probably have difficulty satisfying even this basic requirement.

Why should every enterprise concerned participate directly in the effort necessary to construct the IT infrastructures needed to fulfil all these requirements? Here economies of scale can be achieved by externalization based on managed services.

7

Auto-ID/RFID Infrastructure

7.1 RFID Tags

7.1.1 RFID Tags and Antennas

An RFID tag, which is also called an RFID transponder, consists of an integrated circuit containing a processor and memory combined with an antenna. RFID tags have a wide variety of properties that are significant for specific uses. One of their essential properties is the operating frequency, which with the current state of the technology extends from 125 kHz (low frequency, or LF) to 5.8 GHz (super high frequency (SHF) or microwave), and which has a major effect on the form of the antenna. The following description relates primarily to tags with an operating frequency of 13.56 MHz in the high-frequency (HF) band (Figures 7.1 and 7.2) or a frequency in the range 860–960 MHz (in the ultrahigh-frequency (UHF) band; Figure 7.3). The UHF band is preferred by EPCglobal because it has a larger reading range, which is important for logistics applications.

HF tags are primarily used for identification of individual parts (items). By contrast, UHF tags are primarily used for logistics purposes. Both types of tags can be manufactured in large quantities as disposable transponders. These tags are passive, which means that they cannot transmit signals on their own. Active tags, which have their own source of power, are described in Section 7.1.3.

The principal performance parameters of RFID tags are:

- reading range;
- data transmission rate;
- suitability for bulk reading;
- sensitivity to surrounding objects.

RFID for the Optimization of Business Processes   Wolf-Ruediger Hansen and Frank Gillert
© 2008 John Wiley & Sons, Ltd

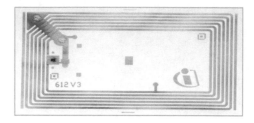

Figure 7.1   RFID tag inlay with coil antennas for 13.56 MHz (source: Infineon)

Figure 7.2   RFID button tag (HF) on an Airbus replacement part (source: Kortenburg)

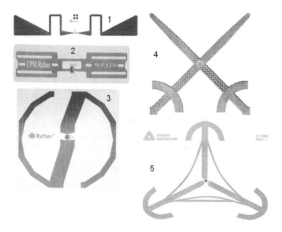

Figure 7.3   RFID tag inlays with dipole antennas for 868 MHz (UHF) (sources: Avery Dennison (1, 5); UMP Relate (2, 3); KSW (4))

The reading range, ability to withstand environmental influences and bulk suitability of RFID tags depend on the operating frequency, the field strength (power) of the reader antenna, the orientation of the tag relative to the reader antenna, and the shape and size of the tag antenna. The frequency and accompanying transmission protocol (including the anticollision algorithm) are the decisive factors for the data transmission rate and reading speed in bulk situations. A bulk situation is always present when several objects with RFID tags pass through an antenna field at the same time, such as on a pallet.

The electromagnetic field between the reader antenna and the tag antenna is called the 'air interface', and the rules that govern data exchange are called the 'air interface protocol' (see Section 7.2). All of this is addressed by the EPCglobal Generation 2 standard.

The manufacturing precursors of RFID tags are called inlays. An inlay consists of a chip and an antenna attached to a carrier film. For use, the inlay can be encased in a hard plastic body to produce a smart card. Alternatively, the inlay can be glued to the back of a label attached to a product, case or pallet and marked with text or a barcode on the front surface.

An inlay for passive RFID tags with an operating frequency of 13.56 MHz is shown in Figure 7.1. At this frequency, the signal is transferred by magnetic fields, so the antenna takes the form of a coil as in a transformer. Figure 7.2 shows another example of a HF tag. It is shaped like a button with a thickness of 2 mm and a diameter of 8 mm. A chip and an antenna are hidden inside the package. This plastic-encapsulated form of tag can be used to identify technical (metallic) replacement parts, for example. In the figure, the tag is shown attached next to a nameplate. The reading range is of the order of millimetres, which is adequate for the intended purpose.

Figure 7.3 shows inlays for use in the 860–960 MHz band. Here the signal is transferred by electromagnetic waves, so the antenna is formed as a dipole as shown in examples 1 and 2. The round form (example 3), crossed swords (example 4) and tripole (example 5) are variants of dipoles. This wide variety of forms is the result of empirical experiments conducted by manufacturers in order to achieve the best possible reception between RFID tags and reader antennas under specific operational conditions for logistics uses and other application areas.

Transponders can also be classified in terms of their storage capacity and whether the stored data can be modified (rewritten). There are four types in this regard:

- **Single-bit, read-only tag**. The simplest type of tag stores exactly one bit and does not contain a chip. These '1-bit tags' are attached to retail products for protection against shoplifting (electronic article surveillance, EAS). After the payment transaction, they are removed at the point of sale. If they remain attached to the goods, they are detected by an antenna at the exit and trigger an alarm. Almost every consumer has already experienced this. These tags do not say anything about the tagged item.

- **Read-only tag with a unique item identifier (UII).** These tags are assigned unique multibit serial numbers during manufacturing. These numbers are used for identification and cannot be modified. If the serial number is used to identify the object to which the tag is attached, a database is used to associate the item number with the tag ID. These are the most economical tags. They can be reused because of the flexible association method, which reduces the effective price per cycle.

- **WORM (write once, read many) tag**. These tags can be coded one time with nonmodifiable data, which can subsequently be read as often as desired (more than 100 000 times). They are used to store item numbers and serial numbers (such as EPC numbers).

- **Read/write tag.** Read/write tags have individually writeable storage areas where data can be written and modified. User data, handling instructions or process data, can be stored in these tags. The data can also be encrypted. Writeable tags can store data that can later be read by the recipient of the object, such as a packing slip. This function is important if the delivery destination does not have access to IT systems that can provide the necessary data, such as with military shipments to remote areas.

7.1.2   Producing Tags from Chips and Antennas

The central element of an RFID tag is a chip that can store data. It is of the order of a square millimetre in size, and it essentially consists of three parts:

- a radio frequency module for signal generation and obtaining operating power from the electromagnetic field;

- a control unit for processing received commands;

- a storage unit.

The most important parameter is the data storage capacity, which is also called the data width. Here a distinction is made between user data memory (such as 96 bits for EPC-compatible chips) and other memory areas used for status control, such as blocking individual memory cells or a kill command for permanent disabling, which is a fundamental requirement imposed by data protection authorities. The memory can also hold data for data verification, such as a checksum. The largest currently available chip for the UHF band has a capacity of 2048 bits (2 Kb), of which 1728 bits can be defined by the user. Users should bear in mind that the more data that is stored in the tag, the longer it takes to read the data.

The purpose of the antenna in a RFID tag is to couple the chip to the electromagnetic field generated by the RFID reader antenna. In the lower-frequency bands (LF and HF) with short reading distances, the tag is in the near-field region of the reader antenna and magnetic coupling with inductors is used for energy and signal transfer. Tag antennas for use in the HF these bands are thus formed as coils (inductive loops). In the UHF band, the tags are located in the far-field region of the reader antenna in the case of relatively large reading distances, and they need dipole antennas for coupling.

The tag antenna should be as compact as possible and easy to manufacture. Dipole and coil antennas can easily be printed on carrier films, which make it possible to produce inexpensive flat tags. However, this has the disadvantage that the antenna characteristics, and thus communication capability, are strongly influenced by materials in the immediate vicinity – in particular, the object to be identified. For example, a dipole antenna is useless on a metallic surface and has different properties on glass than on paper. Consequently, HF tags with coil antennas are preferable for use on metallic surfaces, although a plastic layer is necessary to isolate the antenna from the metal. A tag can easily be rendered unreadable if the wrong type of antenna is selected.

In the fabrication process, the chip is attached to a substrate (such as a film) and connected to an antenna. This produces the inlay. An inlay is transformed into a tag or transponder by packaging it in a self-adhesive label, a smart card or some other usable form.

An inlay can also be integrated into the object to be identified. In this case, a part of the object can be fashioned as an antenna in order to match the antenna to the object as well as possible. This yields a significant improvement in readability and helps protect against counterfeiting.

Dipole antennas are used in the UHF band. However, very small linear dipoles have unfavourable impedance characteristics and thus cannot be connected to chips without large losses. For this reason, dipoles are often folded into serpentine or fractal structures, which yield good compromises with regard to electrical characteristics. Similar considerations apply to simple coil antennas.

The structure and thickness of the material behind the dipole affects the characteristics of the tag antenna. A large metallic surface located half a wavelength behind a dipole can reflect the reader signal and cause signal cancellation at the tag antenna, with the result that the tag cannot be read by the reader. Similar effects can also be produced by nonconductive materials. Adapting chips and antennas to the inlay and its material environment in the tag are thus critical production parameters. The quality of these parameters cannot be seen from the external form of the tag, so it must be determined by testing.

The choice and design of the tag antenna, as well as matching the antenna to the reader antenna field in which it is intended to be read, are critical factors for the desired read rate, which of course should be as close to 100% as possible. Due to the physical considerations described here, it is not possible to have a single tag form that is ideal for all possible uses.

7.1.3   Active and Passive RFID Tags

RFID tags or transponders are classified as active, semiactive or passive. The distinction is based on their functionality, source of energy, intended use, and manufacturing or procurement cost. The type used for a particular application must be selected according to the intended use.

Active tags

An active tag has an internal battery and can generate radio signals on its own. Active tags have a range of about 3 m to more than 100 m, which can usually be configured to meet specific requirements. Active tags can usually be read and written and configured in various modes. In standard mode, they transmit their ID code at a definable interval, possibly accompanied by other data. They can buffer data (such as sensor data) for purposes such as monitoring temperature histories in cold chains or monitoring operational parameters in technical systems such as aircraft engines.

In order to save energy, active tags can also be put in a quiescent mode, which they can only exit in response to a specific request or on the occurrence of a defined event, such as harmful vibration, deviation from a specified temperature range, hazardous operating conditions, etc.

Active tags are larger and more expensive than passive tags, due to the battery. However, certain applications, such as real-time location of cars in car parks, can only be implemented with active tags due the need for a large transmission range.

Active tags are typically reused repeatedly, for example in production control. Transport items are often fitted with RFID transponders in such environments. Each time a blank body is placed on a carrier, its identification number is stored in the transponder on the carrier so the object can be identified automatically by readers at all times during the individual steps of the production process. Such transponders can also store and provide instructions for performing production operations so the instructions are always available locally without requiring access to a central IT system.

Passive tags

Passive tags draw their energy from the antenna field of an RFID reader. They transmit data by modulating the electromagnetic field of the reader antenna. They are thus smaller and less expensive, have a shorter range (a few metres), and have nearly unlimited service life. Read-only, write-once and rewritable tags are available. The unit price of passive tags varies widely depending on the functionality (less than 15 eurocents in mid-2007 for large-volume purchases), but they are less expensive than active tags (approximately 30 euros or more in mid-2007). Passive tags are typically used in logistics applications by attaching them to pallets or cases. They are usually glued to the backs of barcode labels and thus hidden from sight, and they are discarded with the packaging. They are also attached to parts for lifetime identification as previously mentioned.

Semiactive tags

A semiactive tag has a battery but behaves the same as a passive tag. It only becomes active when it enters an antenna field. The battery is then used to amplify the transmitted signal, which increases the useful range.

7.2   RFID Readers and Antennas

7.2.1   Reader Types

Readers are control devices that contain software for controlling antennas and transferring received data to the next higher level of the IT system. Readers are connected to local area networks (LANs), wireless local area networks (WLANs) or other interfaces in order to transfer data (see Figure 7.4). Readers are connected to antennas that emit electromagnetic fields for communication with RFID tags, which are moved past the antennas individually, on pallets or with other objects. The combined antenna and reader is usually called a 'reader' or 'interrogator'.

There are several types of reader/antenna systems as described below:

- **Gate readers (antenna gate).** The reader antennas are attached to a gate through which pallets can be pushed or transported by a forklift truck or lorry. The critical factor here is reliable identification at large distances with various tag position and orientations. An antenna gate

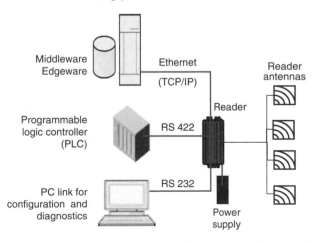

Figure 7.4   Typical reader/antenna connection scheme with one RFID reader and four antennas

typically consists of a reader and four antennas (two on each side of the gate).

- **Compact readers.** Here the antenna and reader are combined in a single housing. This type of reader is a low-cost alternative to a gate reader when the tag distances are relatively short.

- **Vehicle-mounted readers.** For example, readers mounted on forklift trucks.

- **Mobile readers (handheld readers).** Handheld devices for mobile use. These readers can transmit the data read from the tag immediately via a wireless link (GSM or WAN) or store it in built-in memory. In the latter case, the data is transferred after the device is placed in a docking station.

- **RFID-enabled mobile phones (NFC mobile phones).** Mobile telephones can be configured to act as RFID readers as well as RFID tags. This is an economical solution that is significant for consumer applications (such as payment transactions in public transportation systems) as well as industrial applications, such as when a repair technician needs to identify individual parts.

Additional requirements arise from the intended use, such as the protection class of the device. For example, readers and antennas for loading doors must have a relatively large operating temperature range and be protected against dust and moisture.

RFID readers must support suitable interfaces for integration with higher-level software in the RFID system architecture and configuration and diagnostic purposes, such as:

- Ethernet and TCP/IP for WLAN and LAN;

- RS422 or RS232 for direct connection to programmable logic controllers (PLCs);

- GSM, GPRS or UMTS for mobile telecommunications.

Connections to simple control devices are also necessary, such as 24 V digital inputs and outputs for signal lights or barriers that control passage through a gate in a goods receiving or despatching area after the tag data has been checked. Simple programmable logic controller (PLC) setups can also be implemented in this manner. High-level protocols are not yet standardized at this time, which causes additional effort and expense for integrating readers from different manufacturers. A joint initiative on the part of RFID suppliers would be desirable here.

Readers and antennas must above all exhibit an adequate read rate and reading reliability. This depends on parameters such as the number

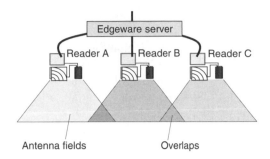

Figure 7.5 Overlapping fields of RFID readers

of antennas and their orientations, the type of tags used, the amount of data stored in the tags, tag placement, etc. Under good conditions, the maximum identification rate is approximately 100 tags per second.

If several readers are used, mutual interference can occur due to overlapping ranges. Figure 7.5 shows a simple example in which the fields of adjacent pairs of antennas overlap. In practice, considerably more complicated interference situations can occur due to reflections from building structures, which also cause read errors. Many of these errors can be detected and corrected in the higher software layer (edgeware) if suitable device management functions are present in this layer.

Data may be preprocessed in the reader, depending on the system configuration. Tag data can be filtered using predefined templates, for example so that only tags attached to pallets are read, but not tags on individual cases on the pallets or the products in the cases.

Another problem that can occur is that the tags on a pallet are read more than once while it is passing through the antenna field. This generates redundant data that must be filtered out before further processing. Good reader software reduces redundant data to a single set that describes the event and is then passed on.

Finally, some readers allow various parameters to be configured under software control, such as setting the radiated power to limit the range. These parameters are stored in the reader.

7.2.2 Reader Antennas

The reader antenna is connected to the reader. It generates the electro-magnetic signals of the antenna field, which form the air interface. In the HF band, the antenna is a coil similar to the tag antenna. It is shaped to achieve the best possible coupling with the antennas of the tags to be read.

There is an especially wide variety of antenna designs for the UHF band, although they are all based on the dipole principle. Highly direc-tional antennas with high gain – in other words, with their maximum power concentrated in the desired direction – are used for large reading distances. As the combination of radiated power and antenna gain is usually restricted by regulatory bodies, the radiated power emitted by the antenna must be configured accordingly. An advantage of highly directional antennas is that their electromagnetic fields can be restricted to areas where the tags to be read are expected to be found. Current European directives stipulate that the antenna power must not exceed 2 W with isotropic (nondirectional) radiation, while levels up to 4 W are allowed in the USA and Australia.

In general, the gain of an antenna is linked to its size by physical laws. The smaller the solid angle of an antenna's radiation pattern, the larger its gain and its physical structure. Highly directional antennas are thus not used in handheld readers. Patch antennas, halfwave dipoles and helical antennas are typically used in such readers.

Relatively large antenna structures can be used with fixed readers. In the UHF band they are usually formed as arrays (see Figure 7.6) in which several small radiating elements are arranged to generate a combined wavefront in the desired direction. The far-field distance of this type of antenna depends on its size. Far-field conditions prevail when the distance to the antenna is large enough that the differences between the path lengths to the individual radiators are negligible relative to the wavelength.

The field is not homogeneous close to the antenna, and tags may not be read if they are located in a region with very low field strength. Particularly in the case of warehouse doors fitted with reader antennas, using highly directional antennas is not always appropriate. Here it is better to use specially designed antennas that generate a measuring field with the best possible homogeneity.

7.2.3 Air Interface

The air interface is the electromagnetic field that forms the link between the reader and the tag. It is an important part of the EPC Generation 2 standard.

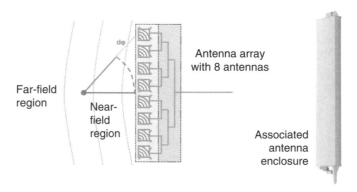

Figure 7.6 An array-type reader antenna and its field regions (source: Kathrein)

The electromagnetic field between the reader and the tag must be modulated to allow them to communicate with each other and exchange data. The air interface specification stipulates the type of modulation and the form of the modulating signal, as well as the structure of the transferred commands, but the actual commands are not part of the air interface.

To enable RFID tags in an open RFID system to be read worldwide, it is essential to have global standards for the air interface and worldwide agreement on the frequencies used for RFID systems. The frequency bands must be harmonized by regulatory authorities and allocated for corresponding use. In the HF band, 13.56 MHz is available worldwide. In the UHF band, 868 MHz is used in Europe, 915 MHz is used in the USA, and other frequencies are used in other regions of the world. Harmonization is not possible due to obstacles such as allocation of frequencies for mobile telecommunication. Consequently, the available UHF frequencies in the 860–960 MHz band must be taken into account for worldwide use.

As readers are transmitting devices, they are only allowed to operate in narrow frequency bands allocated in their region of use. Consequently, passive RFID tags must be able to communicate over the entire 860–960 MHz band if they are to be read worldwide. This imposes high demands on UHF tag antennas and increases their manufacturing cost.

Things are simpler in the HF band because there is a single, internationally available frequency. Here the electromagnetic link is made using relatively simple magnetic coupling. This type of coupling is only slightly affected by nearby objects. The degree of coupling essentially depends on the distance between the reader antenna and the tag antenna and the relative orientations of the two antennas or coils. As a result, tags may be easy to read in certain positions and difficult to read in other positions. In such cases, precise specifications must be generated for aspects such as stacking cases on pallets or attaching tags to cases, and they must be observed in packaging.

In the UHF band with its desirable large reading range, the situation is significantly substantially more complex. Here communication performance is affected significantly by the materials to which the tags are attached and ambient conditions. The entire tag/reader system is affected by other objects behind the tags, and even more by objects in the measuring field (Figure 7.7). The effects include simple attenuation, reflection, refraction and scattering of the electromagnetic waves. Due to the variety of possible interference effects, it is necessary to keep the space between the reader and the tag free of objects during the reading process.

Some exceptions to this rule are thin, nonconductive sheets of materials with low attenuation and small dielectric constants, such as paper, granular substances and expanded polystyrene. It can also happen that tags under a thin covering (such as a tarpaulin) can be read when the

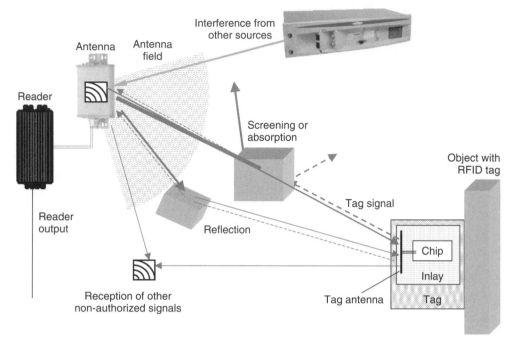

Figure 7.7 Types of signal interference in the reader antenna field (source: Kathrein)

covering is dry, but if the same covering is wet due to rain there is a high probability that the tags cannot be read.

The relative orientation of the reader and tag antennas during the reading process is also critical. The optimum configuration and alignment of the reader antennas, such as on a warehouse doorway, must be determined for each antenna individually and taken into account in packaging (see Figure 7.13).

The reader and the tags communicate via their antennas using a defined command set. The communication protocol is defined such that the reader first sends a command and the chip then replies to it. The command set includes commands for reading and writing data, controlling the anticollision protocol, blocking individual memory cells, and disabling chips ('kill' command).

7.2.4 Bulk Identification

A situation in which several tags appear in an antenna field at the same time is called 'bulk'. For example, this happens if a pallet with individually tagged cases passes through the field. The reader must use a special protocol, such as a bulk identification or anticollision protocol, that serializes the data stream from the tags and avoids collisions so the tag IDs can be recognized. A multitag handling method that enables

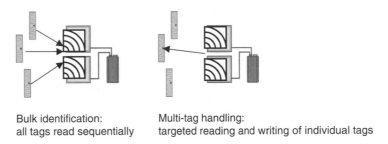

Bulk identification:
all tags read sequentially

Multi-tag handling:
targeted reading and writing of individual tags

Figure 7.8   Bulk identification and multitag handling

selective access to individual tags is used for reading large amounts of data, and especially for write access (Figure 7.8).

The anticollision protocol works as follows. The reader starts by issuing a read command for only one bit of the tag ID, such as Read (1) (see Figure 7.9). All three tags in the figure respond to this command and thus cause a collision. Next, the reader adds a second bit to the ID code with Read (10). Only one of the three tags responds to this command: the tag with ID code 10 111. After this, the identified tag is switched off and the read process is repeated until all of the tags have been identified and no more tags respond. Up to 100 tags per second can be processed using this method.

Tag readability depends strongly on tag materials, structures and positions. For example, it is impossible to read a UHF tag surrounded by metallic objects, liquids or other highly conductive or reflective materials. This means that tags must be attached to the outside of bulk loads, such as on a pallet. They cannot be covered by anything other than paper, expanded polystyrene or similar materials.

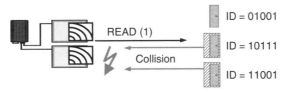

The reader notices a collision due to simultaneous reading of several tags

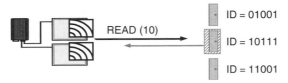

The reader narrows down the tag ID to address a single tag.
All tags are read sequentially.

Figure 7.9   Anticollision protocol for bulk identification

7.3   Near-Field Technology (NFC)

7.3.1   Contactless Smart Cards

Near-field communication (NFC) designates a RFID technology based on inductive coupling with a reading range of a few centimetres. It operates at a frequency of 13.56 MHz with a maximum data transmission rate of 424 kbps. NFC is used in particular with smart cards, electronic passports, loyalty and signature cards and mobile telephones that act like smart cards. It has already been defined by several standards: ISO 18 092, ECMA 340 and ETSI TS 102 190.

In 2004 Nokia, Sony and Philips founded the NFC Forum (www.nfc-forum.org), an industry association intended to promote widespread use of NFC technology. In addition to the founders, its members now include American Express, Atos Origin, Giesecke & Devrient, Master-Card, Microsoft, Motorola, NEC, Panasonic, Samsung, Texas Instruments, Vodafone and Visa.

Manufacturers of mobile telephones participate in the NFC Forum because NFC provides better communication than Bluetooth and WLAN and enables simple, intuitive communication. Instead of using an elaborate pairing process with several distinct steps, the user simply holds a Bluetooth PDA and a suitable mobile phone close together for a short time. The devices identify each other via NFC and open a Bluetooth data link without the usual bandwidth search, device selection, service selection or password transfer. This intuitive form of device communication by simply holding two devices together is typical of NFC.

NFC is expected to be used with smart cards, RFID tags, bank cards, credit cards, entertainment electronics, mobile phones, PCs and other smart objects. Companies such as Visa and MasterCard have already issued NFC credit cards that accelerate wireless payment transactions and enhance their security. There were more than three million NFC smart cards in use worldwide at the end of 2005. The admission tickets for the 2006 football World Cup in Germany were also fitted with NFC chips to prevent theft and counterfeiting and make stadium access faster and more secure.

As an international standard, NFC is intended to be compatible with prior competitive smart card systems from Philips (Mifare) and Sony (FeliCa) and ultimately replace them. Following the example of the WiFi Alliance, the NFC Forum is intended to drive implementation and standardization of NFC technology.

Pilot projects with NFC smart cards have demonstrated relevant economic improvement potential, which according to statements by Munich-based Giesecke & Devrient, a manufacturer of NFC smart cards, can be described as follows:

- Increasing the billing amount per transaction by approximately 10%.

- Reducing point-of-sale (POS) transit time by 8 to 12 seconds relative to cash payment. Quotes:

  - 'Unit sales increased by one percent for every six seconds saved at the drive-in' (McDonalds, USA).

  - 'In every lost second, we lose 1.5 million pounds' (Tesco, UK).

- Increasing frequency of use by 27%.

- Reducing POS transaction time.

- Increasing POS transaction reliability.

- Increasing customer loyalty and customer satisfaction.

7.3.2 Electronic Passports: e-passports

Since November 2005, every German citizen who applies for a passport receives an electronic passport with an integrated RFID tag. Germany was the first European country to take this step. It was not taken entirely voluntarily, but instead was strongly influenced by the demands of the USA to increase security with respect to arriving travellers.

The German e-passport has an integrated IC from Philips (72 Kb SmartMX) or Infineon (64 Kb SLE66CLX641P), both of which are based on the ISO 14443 standard and communicate on the internationally available 13.56 MHz frequency. Philips provides a pure type-A IC in accordance with ISO 14443-A, while Infineon supports both type A and type B. Both ICs are designed for a maximum reading distance of 10 cm (proximity coupling), although experts estimate that the value may be slightly less in actual use. These passports are placed directly on top of an RFID reader antenna for reading. The text data must be read optically before the data in the RFID tag can be read. This provides reliable protection against unauthorized reading of the RFID tag.

Originally, the chip in an e-passport stored only a photo of the holder in addition to the usual passport information. After suitable cameras have been installed at border posts in a few years, a photo taken on the spot can be compared with the stored photo in order to automate personal identification. The next step has already been taken in Germany (November 2007), where the issuing authorities also scan the fingerprints of both index fingers and store them in the e-passport. After this, personal identification cards are also supposed to contain chips with this capability.

In the future, the same RFID tags and biometric security measures will be used to identify travel visas. Accordingly, in the next few years the European Union will establish central fingerprint databases to combat fraudulent entry.

7.3.3   Mobile Telephones with Smart Card Properties

The world's first RFID mobile telephone with NFC capability was announced by Nokia in February 2005 as an enhanced version (model 3220) of an existing model. It is based on the ISO 14 443-A standard and is intended to support all functions supported by smart cards, such as payment transactions, ticketing, etc. With an NFC mobile phone, the user can initiate a transaction at a point of sale or entrance to a sports facility by touching the phone to a marking that indicates the location of an RFID tag or antenna. The credit card data and ticket information are stored securely in the smart card chip. The advantage of a telephone is that it can transmit as well as receive via the NFC link. Peer-to-peer communication is also supported, which for example means that two mobile phones can exchange data with each other directly.

Immediately after the announcement of the first NFC mobile telephone, Visa and MasterCard stated that they welcomed it as complement to their credit cards. NFC telephones are already being used successfully in the public transportation system of Hanau near Frankfurt, Germany. Besides ticketing, there are numerous other potential applications for NFC in the consumer and industrial sectors, for example when repair technicians need to identify individual parts and access more information about these parts in the transponder memory or central databases (see Sections 10.5 and 10.6).

7.4   Prerequisites for RFID Infrastructure Implementation[1]

A structured approach is essential for introducing RFID technology[1] in an enterprise. Such an approach is described here from the technical and process perspectives. This description is based on experience gained from many research projects and industrial projects carried out by the Logistics Department (FLog) of the University of Dortmund under the leadership of Professor Rolf Jansen PhD, and in particular experience from a current project focusing on container management in the automotive industry. However, it is equally relevant to other application areas.

7.4.1   Staged Approach

The process of implementing a RFID infrastructure can be divided into four stages. The necessary stages are requirements analysis, laboratory studies, system design, and implementation in a pilot system. The result of the requirements analysis is information that forms the basis for discussing alternative solutions and making decisions regarding preliminary selection

---

[1] Contributed by Jan Hustadt, Logistics Department, University of Dortmund (Director: Professor Rolf Jansen). www.FLog.mb.uni/Dortmund.de

of the system components. The purpose of the laboratory studies is to provide basic proof of technical feasibility. The first two stages form the basis for the system design. The RFID system and its infrastructure are first described in sufficient detail at the theoretical level in the system design stage and then implemented in the pilot system. The process ends with the rollout decision. A combination of linear and iterative procedures is shown in Figure 7.10. Generally speaking, the cost of making changes and refining the design is inversely proportional to the level of detail in the preparatory stages and the size of the feedback steps.

7.4.2   Stage 1: Requirements Analysis

Requirements analysis forms the basis for the following stages and is thus critical for the success of the project, although it is not always given the attention it deserves. It essentially consists of process analysis, environment analysis, object analysis and IT infrastructure analysis. However, these aspects need not be viewed entirely separately, and in fact they should not be viewed separately. They actually represent different views of the same situation, as can also be seen in particular from the gradual merging of the real world and the virtual world of IT systems fostered by RFID in the sense of the internet of things or ubiquitous computing.

Ideally, requirements analysis should be preceded by an analysis of the current situation and its weaknesses. In addition, in most cases there are prior solutions for RFID systems, so it is generally necessary to answer questions regarding integration into existing systems and transition from one system to another one. Another question is whether the necessary RFID-driven changes are actually possible and if so, how. These questions also form the subject of the following reflections.

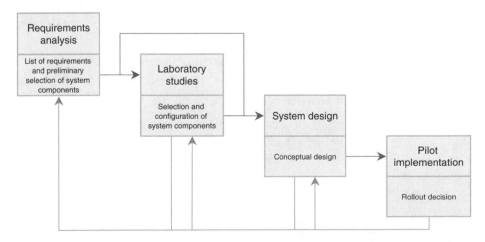

Figure 7.10   Stage model for RFID implementation (source: FLog, University of Dortmund)

Process analysis

Process analysis is based on models that represent an initially preliminary target concept, the material and information flows of the business process concerned, and its necessary activities, such as production or administration activities. At this stage, fundamental attention should also be given to aspects such as system management, ease of use and reliability, as outlined below in the context of system design. Sufficiently detailed descriptions are necessary, in part so that the questions listed below can be answered. They are intended to convey a small insight into the highly complex process of implementing RFID systems.

- Where are the reading locations, and when is data read?
- When and where is the data written to the transponders?
- Are fixed and/or portable devices to be used?
- When, where and how are the transponders attached to the objects?
- Will the transponders be used several times?
- If yes: when, where and how are the transponders removed from the objects?
- What reading and writing distances are necessary?
- Is there a need to maintain safe distances?
- Will the objects be moving during the reading and writing processes?
- If so, how fast?
- What types of conveying equipment will be used?
- How many transponders must be acquired concurrently?
- Is the functional scope of the RFID system expanded or extended by sensors?
- Are sensors and actuators used to control the system?

Environment analysis

RFID systems are constantly or temporarily exposed to a large number of external influences and loads arising in part from their environment and in part from the application. Without going into the details of the underlying physical laws, we can say that this usually has a negative impact on communication between transponders and readers and on the robustness of individual system components or their connections.

For example, the coupling of the transponder and reader antennas for communication is based on alternating magnetic fields (at frequencies below 13.56 MHz) or electromagnetic waves (at 433 MHz,

860–960 MHz or 2.45 GHz).[2] The attainable data transmission rate is directly related to the frequency. The higher the frequency, the faster the data can be transmitted, which is a significant factor in connection with anticollision methods (for bulk reading), among other things. Communication disturbances occur for various reasons, but generally speaking it can be said that electromagnetic waves are reflected by certain surfaces, which results in interference effects that cause signals to be reinforced or cancelled ('signal holes'). When electromagnetic waves encounter electrically conductive surfaces, they induce eddy currents that dissipate energy. With alternating fields, eddy currents can also be induced in metal structures such as the framework of a continuous conveyor, and under some conditions these eddy currents can unintentionally enlarge the effective acquisition range. Materials with large dielectric constants, such as water, absorb energy from the antenna field, especially at high frequencies. This produces undesirable effects, which in essence can be summarized as a reduction in the effective range for reading and writing processes, acquisition of transponders outside the defined acquisition area, and holes in the acquisition area.

In any case, it is necessary to ensure functionality, which means successful communication between transponders and read/write devices (readers), under all relevant environmental conditions. This requires making suitable adjustments, such as to the infrastructure and the system, which can only be done effectively if the relevant factors are known. Table 7.1 provides an overview.

Table 7.1 Summary of external influences and loads

Electromagnetic Fields	Mechanical Loads	Chemical Loads	Thermal Loads	Weather Factors
Reflective or electrically conductive surfaces, absorbent materials, total RF spectrum (interference frequencies), electrostatic discharge, etc.	Shocks, vibrations, pressure, friction, shear forces, etc.	Oils, cleaning products, lubricants, acids, alkalis, surfactants, solvents, etc.	Operating and storage temperatures (minimum and maximum)	Rain, fog, atmospheric humidity in general, frost, ice, solar irradiation, salty sea air, etc.

---

[2] See the ISO/IEC 18 000 standard for more information.

Object analysis

Object analysis encompasses all relevant aspects related to the objects to be identified. From a logistics perspective, at present these objects are primarily products, packaging and, in particular, load carriers, such as used in connection with container management, and load units, but other types of objects are possible. The following questions must be answered:

- Which objects are to be fitted with transponders?
- What are the sizes of the individual objects?
- Are the objects filled with contents, and what are the contents?
- Are the objects always observed in the same state? (Reading in full and/or empty states, rigid/nesting/folding containers, etc.)
- Are individual objects or groups of objects to be observed?
- In the latter case, are the groups homogeneous or heterogeneous?
- How many objects belong to a group?
- How are the objects of a group arranged?
- Are there any objects that are fully surrounded by other objects?
- Are there objects enclosed by other objects?
- Are the objects of a group always observed in the same arrangement?
- Are the objects enclosed by materials that protect or secure the objects?
- Are there any specifications for the size, position and/or attachment of transponders?
- What is the substrate material for attaching the transponders?
- Can more than one functionally equivalent transponder be used with each object?
- In case of a group, is a representative of the group sufficient to identify the group?

IT infrastructure analysis

The primary focus of IT infrastructure analysis is the data stored and processed in the system, as well as the related questions:

- Which data will be stored in the transponders?
- How much storage space must be reserved for this?

- Is it necessary to associate data in the system with data in the transponders?

- What type of data storage and management is desired (centralized, decentralized or hybrid)?

- How much data must be transferred in the read and write processes?

- Will the data in the transponders be modified? If so, how often?

- Are there preferred frequency bands or transmission protocols?

- Is it necessary to encrypt data, and are access privileges necessary?

- Which data and privacy protection regulations must be obeyed?

In addition, the existing IT system landscape of the enterprise must be analysed so the requirements for the edgeware/middleware and underlying systems can be derived in conjunction with the results of the process analysis. This step is necessary because individual producers take different approaches to connecting and integrating RFID systems. Consequently, a decision for or against a particular system affects aspects such as filtering and preprocessing of data from the readers or integration of devices that enhance ease of use. Other considerations such as existing network structures, interfaces associated with such structures, available data processing capacity, the amount of data to be processed, and provision of operating power (which at first glance may seem trivial) are also important. Space does not permit a more detailed discussion of this topic, so we must omit it here.

Another aspect that must be considered is that exchanging objects across enterprise boundaries presumes at least the existence of RFID agreements, or better yet standards, regarding the transponders to be used and the associated air interfaces and data structures.

7.4.3   Stage 2: Laboratory Studies

In principle, the field strength of signals emitted from an antenna, and thus the range within which read and write operations are theoretically possible, can be calculated at any point in space using Maxwell's equations together with various boundary conditions. However, calculations of this sort are highly time-consuming and labour-intensive, particularly with regard to constantly changing conditions in actual operation, so an empirical approach focused on processes and practical tests is a practical option. It also evidently eliminates the need for facilities such as screened rooms and the associated scientific rigour.

The objective of laboratory studies is to provide basic proof of the technical feasibility of a planned use of RFID technology in a specific application scenario. In this connection, a multistage procedure focusing on the performance of the RFID system and the robustness of the individual system components and their connections has proven its worth in practice. Laboratory tests offer the following advantages relative to direct on-site use:

- They are performed in a known environment in which all conceivable ambient conditions, ranging from low-noise, low-reflection conditions to simulated actual conditions, can be generated. This allows a stepwise (modular) approach as necessary to be used as needed to achieve a functional RFID infrastructure, with the knowledge that the ambient conditions can be controlled in a reproducible manner, unlike on-site use in an enterprise.

- Objective assessment is possible due to optimum conditions.

- The resources of the enterprise are not burdened unnecessarily, and its operation is not disturbed.

Among other options, these studies can be performed by the FLog LogIDLab® in Dortmund, Germany, assisted by the resources and capabilities of the accredited, affiliated PackLab®, using an extensive, standardized and adaptable spectrum of test and verification programmes listed in Figure 7.11. In this regard, an open standard for RFID test methods that is not bound to membership in a particular organization or the

Testing the ability to withstand specific environmental factors	Testing performance in low-noise, low-reflection environment and with respect to specific environmental factors	Supplementary determination of electromagnetic properties
Mechanical robustness (static and dynamic), ability to withstand weather conditions, thermal robustness, chemical resistance, etc.	Attachment substrate (plastic, metal, corrugated cardboard, etc.), material penetration reading range, bulk capability acquisition range of various antenna systems, etc.	Electrical and magnetic field strength, Q factor of inductive systems, resonance frequency and bandwidth, minimum magnetic field density at various read and write frequencies, spectrogram, etc.

Figure 7.11 LogIDLab® test and verification programmes for RFID systems

like is available in the form of ISO/IEC TR 18046. With the assistance of FLog, the VDI (Verein Deutscher Ingenieure/Association of German Engineers) issued an acceptance procedure for verifying the performance of RFID systems as part of its 4472 guideline.

Performance

Studies of the performance of RFID systems have two principal focuses: determining suitable locations for attaching transponders and determining configurations for antennas and downstream readers. one method that can be used to tune this system is a two-stage procedure consisting of static tests followed by dynamic validation. In this regard, preliminary selection of system components based on the requirements analysis helps minimize the scope and cost of the tests. However, it should be noted that performance differences can also result from differences in the software and firmware of the readers. Static tests give a very good indication of the capabilities and limitations of an RFID system. These tests essentially address reading range, position and orientation sensitivity, and attachment substrates.

The reading range is defined as the maximum usable distance between the transponder and the reader antenna along a straight line extending from the centre of the antenna and perpendicular to the antenna portal. Due to the three-dimensional nature of the effective antenna field, other distances are also examined here, in particular on two mutually perpendicular planes – one horizontal and the other vertical – intersecting the above-mentioned line over its entire length. To determine the reading range, the transponder is initially placed at the centre of a neutral or typical substrate or a substrate that best matches the anticipated use and oriented parallel to the reader antenna, which is the orientation that yields the best results.

To investigate orientation dependence, the approach described above must be modified by varying the alignment of the transponder relative to the reader antenna. Figure 7.12 shows measured results of range and orientation dependence tests for a square antenna (left side) and a rectangular antenna (right side) with the maximum allowable antenna power of 2 W ERP. Both transponders operate in the UHF band (868 MHz) in accordance with the Generation 1 EPCglobal standard (Specification 1.19). Their shapes are show in Figure 7.12. Three different orientations were tested in a large, open outdoor area in order to eliminate interference effects:

- transponders and antenna parallel (upper charts);

- transponders vertical and perpendicular to the antenna (middle charts);

- transponders horizontal and perpendicular to the antenna (lower charts).

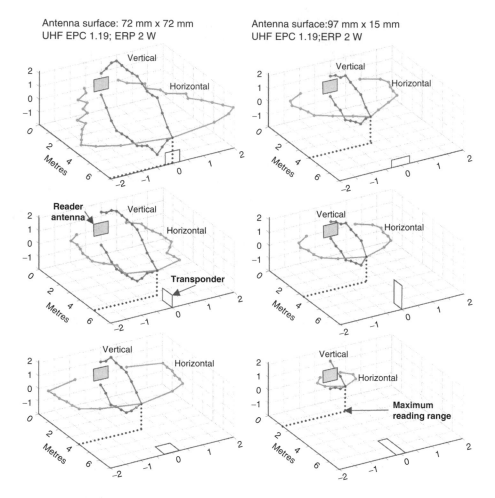

Figure 7.12   Reading range of UHF transponders in three different orientations for square (left) and rectangular (right) formats (source: FLog, University of Dortmund)

The charts show the antenna lobes, which means the experimentally determined maximum ranges in the vertical and horizontal planes for the two UHF transponders. It can be seen that the maximum reading range is obtained for both transponders when they are parallel to the reader antenna. The reading range decreases significantly with a perpendicular orientation. With the rectangular transponder, the range decreases dramatically when the transponder is horizontal because the transponder has the least 'visibility' relative to the reader in this orientation. The maximum reading ranges measured by FLog, which means the tips of the antenna lobes, are indicated by the dotted lines in Figure 7.12.

These facts must be taken into account when configuring transponders and reader antennas. Even if transponders are correctly attached to objects such as pallets or cases, simply turning an object by 90° can make the transponder unreadable.

It may also be worthwhile to perform the test with a variety of substrates in addition to various orientations if various objects made from different materials must be considered. The general impact of various materials on reading range is shown in Figure 7.13 for HF transponders and Figure 7.14 for UHF transponders. Besides the studied materials, other candidates for fundamental studies include steel, aluminium, sheet glass, acrylic plastic, corrugated cardboard, polypropylene, polycarbonate, other plastics, foams, films, chipboard, plywood, and so on, in part in both dry and moist states.

In order to determine the best position for a transponder on a particular object, it is helpful to combine the insights obtained from the static tests with the results of additional studies, such as investigation of the resonant frequency of the transponder as a function of possible attachment positions, and the accumulated knowledge of institutions such as FLog.

After this, the defined positions must be checked and confirmed by dynamic validation. Questions regarding the optimum antenna configuration must also be answered, in particular the number of antennas and their arrangement, including spacings and orientations. In addition, if necessary, aspects such as antenna radiated power, antenna drive and inclusion of sensors and the like must be studied. This requires simulation

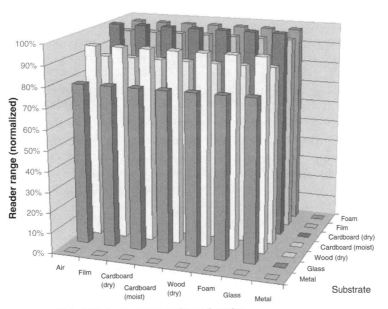

With the exception of metal, the range is not affected by the various materials with the same substrate.

Application on a metal substrate is not possible without adaptations.

In comparison with dry materials, moist materials provide only slightly poorer reading characteristics.

Although the substrate material has a larger effect on the reading range than the material between the transponder and the reader, its effect is nevertheless small.

Antenna surface: 80 mm x 50 mm, ISO 15693

Figure 7.13 Effect of various materials on the reading range of an HF system (source: FLog, University of Dortmund)

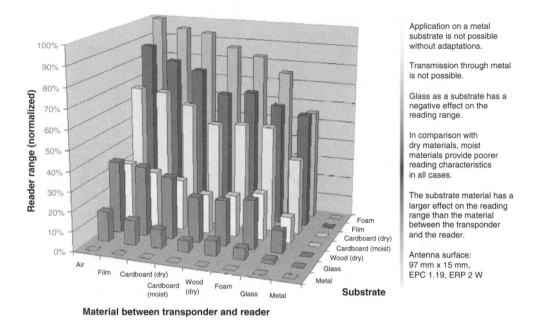

Figure 7.14   Effect of various materials on the reading range of a UHF system (source: FLog, University of Dortmund)

of the anticipated use scenario in order to model the dynamic behaviour of the conveying equipment used in actual operation or comparable equipment. If there are special external factors, they must also be simulated at a suitable place.

If bulk reading capability is significant (which means the ability to communicate quasiconcurrently with several transponders in the antenna field), this can be investigated either before or during dynamic validation by moving various numbers of transponders with fixed spacing past an antenna at a walking pace. The results can be evaluated in each case by comparing the number of transponders for which communication was successful with the total number of transponders available to the antenna. Here it is necessary to distinguish between one-, two- and three-dimensional arrangements, and the transponder spacing must not be less than a certain value. The point of reference is always the centre of a set of transponders. This test can also be performed at various speeds to obtain information about another practically relevant parameter that correlates with bulk reading capability, as illustrated in Figure 7.15. This figure clearly shows that a 100% (or nearly 100%) read rate, which is often quoted in relevant publications and demanded by users, cannot be taken for granted in practice and cannot be achieved without suitable technical and/or process adjustments.

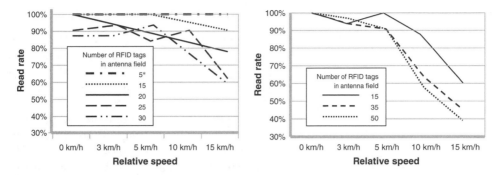

Figure 7.15 Read rate as a function of relative speed. Left: 13.56 MHz HF tags; right: 868 MHz UHF tags (source: FLog, University of Dortmund)

Robustness

Besides the performance of RFID systems, the robustness of individual components and their connections (which relates primarily to the attachment of transponders to objects) with respect to external influences must not be ignored. This also applies to the objects, since the transponders can also be affected if parts are damaged or are at risk of falling off. Damage to load carriers, which under certain conditions can also affect the functionality of transponders, is by no means unusual in everyday operation.

In addition, RFID systems in a production environment are exposed to mechanical, thermal and chemical stresses (such as in painting lines in the automotive industry). It should also not be assumed that ideal climatic conditions can be found in the cargo spaces of various types of transport vehicles. This applies in particular to air cargo (for example), and less so to short-distance road transport. Transponders for identification of aircraft parts must fulfil very stringent requirements with regard to durability under unfavourable climatic conditions with large temperature fluctuations and a wide temperature range, as well as additional mechanical stresses such as vibration and shock.

Aside from this, unfavourable conditions can cause adhesives to fail with the result that transponders simply fall off objects. In the case of courier, export or parcel services and handling of shipping units, there is no guarantee that transponders attached to the outside will not be damaged by impacts from the corners or edges of other shipping units. Aside from transponders, readers and other devices are also exposed to vibration, shocks and other influences if they are attached to a floor conveyor vehicle or integrated into a container, for example as part of an onboard unit. All of this – and more – can be suitably simulated and investigated in an ideal manner in facilities such as the FLog PackLab®. The tests and studies that are actually necessary must be determined for each case individually.

7.4.4   Stages 3 and 4: System Design and Pilot Implementation

The system design combines the results of the requirements analysis and the laboratory studies in a manner that generates a comprehensive, definitive high-level design using RFID technology. At least in theory, the high-level design is then converted into a practical design implemented on site in a pilot plant. The pilot system encompasses all aspects that are significant for making a rollout decision. For example, the pilot system approach means that instead of immediately fitting all goods receiving doors of a large distribution centre with RFID technology, only part of the centre is fitted with a system that addresses all the technical and process aspects, and this is done step by step. Here the considerations and results of the process analysis are applied to the material and information flows, among other things, and incorporated into the target process as part of the system design. In addition, the elements of the requirements analysis (such as the data to be stored) are regarded as fixed quantities that do not need any further detailing.

A complete analysis of an RFID system requires considering not only the transponders and RFID readers, but also the elements of the IT system that receive and process the RFID data: the middleware, the databases and the applications. This will clearly indicate whether adequate information about the hardware infrastructure and major portions of the software infrastructure is already available. However, it must be noted that a reader can only control a few antennas and it is not possible to operate an unlimited number of readers from a single computer via edgeware and middleware. Aside from this, the key task of system design, as already mentioned, is combining the known components with consideration to software implementation, system management, ease of use and reliability. Here we can touch briefly on selected aspects.

Aside from the aforementioned facets and obvious elements, software implementation encompasses (among other things) synchronization concepts and consistency analyses of the data available in the system. In addition, the tasks and functions of the edgeware and middleware must be defined and embedded in the software. This includes the events, such as passage of something through a warehouse door. Here it is necessary to define certain time parameters, since the reader will identify the transponders several times in such a process but there is only one event of interest for further processing.

These considerations are linked to arriving at a system management scheme. For example, in a UHF system that complies with European regulations it is not possible to operate several readers in parallel if the number of readers exceeds a certain value. According to the regulations, reader synchronization functions in particular and reader control functions in general must be provided, for example by using sensors and/or actuators such as light barriers, motion detectors or switches. In addition, software extension protocols (for updates, upgrades, additional components, etc.) must be integrated. At present, updating reader firmware is still a very laborious process, which becomes even more significant with an increasing number of readers.

A related topic is ease of use. The objective here is to make it easier for users (employees, etc.) to deal with the new system, which automates processes and is fundamentally invisible. Some possibilities here are acoustic signals, visual indicators such as traffic lights, and screen displays. Other user aids include the ongoing use of proven internal part markings, such as text labels.

An issue that will become increasingly important in the future is the reliability and/or stability of RFID systems. The following questions (among others) must be answered:

- How can a consistently high level of transport quality be ensured?
- Can transponders from other manufacturers be used?
- How are malfunctions and component failures detected?

7.4.5 Summary

This section shows that configuring an RFID system with regard to selecting RFID tags, tag placement on objects and antenna configuration is a complicated process. Simple solution templates are still a long way off, in part due to the specific conditions of objects and antennas or antenna environments. Self-configuring systems can thus not be expected in the foreseeable future. Given the staged process described here, a large amount of integration effort will be necessary for future systems, and the suitability of potential technologies must be determined on a case-by-case basis. Experienced independent institutions, such as FLog at the University of Dortmund, are thus attractive and valuable partners for performing objective analysis and formulating specific action recommendations.

7.5 Global Positioning with GPS/GPRS

In order to recognize RFID tags, it is necessary to have suitable RFID antennas. For this reason, the doors of distribution centres and goods receiving and despatching warehouses in the trade sector will ultimately be fitted with RFID antennas. But supply chains span the globe, and goods are transported by road, water and air. It is clearly unlikely that enough RFID antennas can be placed along these routes to provide comprehensive monitoring of global supply chains. GPS devices can close this gap and enable monitoring of containers or other transport media on their worldwide routes using satellite positioning.

Every position in the world can be determined by measuring distances to a group of satellites. Special satellites transmit their positions to the ground at regular intervals so these distances can be measured. At present, this service is provided by the American GPS satellites and the Russian GLONASS satellites. The European Galileo system is planned to be available after 2012. Its first satellite for testing purposes was launched into orbit in late 2005.

Position determination works on the principle of measuring the propagation times of the satellite signals in order to determine the distances to the reference points. A position can be calculated from the distances to at least three satellites. This is similar in principle to a cross bearing in sailing. The accuracy of the satellite method is between 2 and 20 m.

The accuracy can be improved by using electronic services[3] that considerably refine the position data by comparing them with fixed reference points. Services of this sort are used for applications that depend on exact geographic position information, such as laying gas pipelines. Communication with the reference service is via GSM or GPRS. The two satellite systems (GPS and GLONASS) can also be used together to achieve improved accuracy. The Galileo system can be included in the process once it is operationally available.

Aviators, sailors and motorists are familiar with GPS positioning systems. They are integrated into the navigation systems of aircraft, ships and motor vehicles. In the logistics area, they have proven their worth for fleet management of trucks. The rental bicycles of German railway operator Deutsche Bahn are also located using GPS and released for use via GSM.

However, international logistics involves containers and other load carriers that do not have their own sources of electrical energy, unlike motor vehicles. Consequently, the GPS devices must be fitted with batteries, which at present is a critical problem due to insufficient battery life. GSM or GPRS links are also necessary for transmitting position data to the monitoring IT systems. This means that the GPS device must always be linked to a mobile telephone.

In the simplest case, a suitably equipped container transmits its position at regular intervals so the monitoring IT system can continually recognize and track its position in the supply chain. This is a typical example of tracking and tracing, which involves monitoring schedules, detecting deviations from planned routes and times, and obtaining indications of fraudulent diversions. Route positions and times are stored to allow routes to be verified afterwards whenever desired.

This sort of positioning is also an attractive option for semitrailers, since the positioning system of the tractor is of little use if the trailer is decoupled and then coupled to a different tractor. Quite a few loads have

---

[3] See, for example, the ascos service of e.on (http://ascos.eon-ruhrgas.com).

already been diverted to untraceable routes this way. If a container with a positioning device is loaded on a lorry, the container can be monitored continuously.

7.5.1   Monitoring Sensor Data

As we have seen, a GPS system must always be connected to a second component: a GSM communication device. A third element – sensor electronics – is added if it is necessary to not only determine the geographic position of the container, but also measure and store data for other parameters such as:

- the temperature of a cold container (to secure a cold chain);

- vibrations that could damage the load;

- unauthorized breaking of an electronic seal (e-seal), which is an indication of tampering with the load.

Devices linked to GPS systems are already being used in motor vehicles to record and transmit operational data. This supports remote diagnosis and preventive maintenance.

7.5.2   Functions for Sea Transport

Several technical obstacles must be overcome to enable widespread commercial use of the container positioning systems described here.

Battery lifetime

Batteries should have service lives corresponding to the service lives of the containers, or at least their maintenance intervals, which means 2 to 8 years. This is still difficult at present. However, rechargeable battery technology is evolving and the power consumption of positioning devices is constantly decreasing.

Additional battery life can also be achieved by implementing effective power management in the positioning device software so the tags are only enabled when necessary. For example, route parameters can be stored in the device to invoke transmission of routine messages at important control points instead of periodically at unnecessarily short intervals, and naturally to trigger alert messages in case of unforeseen events.

Communication outside of GSM reception areas

GSM communication is tied to terrestrial antennas and only available at sea in coastal areas. However, it is essential for shipping customers to receive messages during sea transport as well, for example if the specified temperature of a cold container is not maintained or an electronic seal is broken. Temperature deviations cause goods to spoil and render them worthless for trade. This means that a new shipment must be ordered immediately to maintain the planned availability.

The specifications of the new Galileo system suggested that it would provide a separate communication channel that could be used at sea. However, this is not so because this channel is intended for emergency use only and does not have nearly enough capacity for additional message traffic for tracking and tracing.

A possible solution would be to provide GSM infrastructures on ships. This would allow GSM calls from containers to be received and conveyed transparently via satellite communication. Suitable products are commercially available [Tria04].[4]

Communication from 'buried' locations in ships

If a GPS system inside a ship is covered by numerous other containers, it loses contact with the satellites. In this case, even the supplementary Galileo channel would be of no help. There are two possible solutions to this problem.

First, we see that methods for direct intersystem communication are being developed by research groups and industrial organizations under designations such as agent technology (see section 6.2, 'Software Agents'), ubiquitous computing and meshed networks. It is thus conceivable that messages from containers at the bottom of the stack could be transferred to containers at the top for transmission if most of the containers on the ship were fitted with agent-enabled GPS devices. Another conceivable option is to provide a communication infrastructure on board the ship for use by all the containers, similar to the present situation at airports and in railway carriages.

7.5.3   Smart Containers with RFID and GPS

There is every reason to believe that container tracking and tracing will become a reality. With regard to the complementary significance of GPS and RFID, we can mention the RFID Smart Box,[5] which was first exhibited in Magdeburg, Germany, in 2005 by the Fraunhofer Institute for Factory Operation and Automation (IFF). It contains a combined GPS/RFID system for secure transport of valuable goods such as mobile telephones (see Figure 7.16).

---

[4] See, for example, the BGAN service of Inmarsat Global Ltd, London (www.inmarsat.com).

[5] See Logmotionlab ('laboratory for mobile logistics objects') at www.logmotionlab.de.

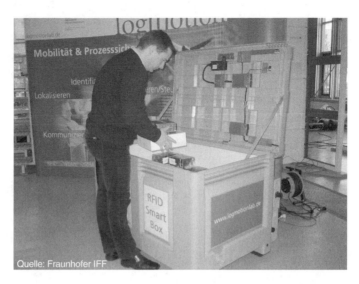

Figure 7.16   A container with an e-seal, RFID antenna and GPS positioning system (source: Fraunhofer IFF)

The box is approximately hip-high and the size of a large chest freezer. The lid of the box has an electronic seal and can only be opened with a smart card containing an RFID tag and authorization data. An RFID antenna is integrated in the box to recognize goods that are put into or removed from the box. The transported cases are fitted with RFID tags for this purpose. The geographic position of the box is determined using GPS. In a demonstration, the box was transported through the city and its contents were distributed to various customers. A central IT system constantly recognized where the delivery van stopped, who opened the Smart Box and which goods were put into or removed from the box at each location [Enai05].

In cooperation with Maersk Logistics in Denmark, one of the largest logistics enterprises in the world, IBM is developing a comprehensive software system for global tracking and tracing of containers under the name Secure Trade Lane. The project scope includes a tamper-proof electronics box dubbed 'TREC' (Tamper Resistant Embedded Controller), which is installed in a container and transmits data to the system. This system is intended to create high visibility for global supply chains and fulfil regulatory (homeland security) and security (electronic seal) requirements [Rindle2607]. It can also fulfil the requirements of customs authorities if it is implemented in compliance with the framework agreement of the Brussels-based World Customs Organization (WCO; www.wcoomd.org).

A system that combines GPS and RFID technology to help potential purchasers find cars in which they are interested on the sales lot of a car dealer is also described in Chapter 10.

GPS technology has already proven to be a worthy complement to RFID, and this trend can be expected to continue. The obstacles with regard to battery life and communication outside the range of terrestrial RFID antennas, although still present, appear to be surmountable.

8

Consumer Protection and Data Protection in RFID Applications

8.1 RFID and Data Protection

Attention must be given to data protection and consumer protection requirements in RFID applications. The primary requirement is to ensure the right of citizens to determine how their personal data is used, which is sometimes called 'informational self-determination'. Objects with RFID tags will often end up in the hands of individual persons and consumers at the end of the supply chain. In order to avoid surprises, false insinuations and unjustified fears, all business processes that are associated with RFID technology should be evaluated with regard to their relevance to data protection. Preventive information is urgently necessary in situations where consumers come in contact with RFID.

This chapter presents an abridged version of the *RFID und Datenschutz* ('RFID and Data Protection') guideline of the EICAR RFID Task Force [Eicar2006], which is intended for enterprises that wish to use RFID technologies. It provides information and recommendations on data protection aspects related to RFID, particularly in the consumer domain, where in the future it will be possible to link logistics data in RFID chips to personal data.

Naturally, another important consideration is ensuring the security of RFID systems. With regard to this topic, we refer the reader to the study of the German Federal Office for Information Security (BSI) [BSI2004].

Experience shows that the impact of RFID tags on consumers must be assessed in legal as well as informatic terms. Many enterprises that pioneered the use RFID tags in the early days of technological euphoria have had painful experiences as a result of inadequate communication. Consequently, groups of persons who come into contact with RFID tags as consumers or customers – especially in shops and supermarkets – must

RFID for the Optimization of Business Processes   Wolf-Ruediger Hansen and Frank Gillert
© 2008 John Wiley & Sons, Ltd

always be informed about RFID-related measures comprehensively and with an eye to future developments. Good public relations work is indispensable here.

In a study by the University of St Gallen, it was found that RFID can only be successful in practice if the interests of data protection are taken into account. EICAR has thus produced a guideline with the aim of providing practical advice to enterprises that wish to use RFID in areas directly related to consumers while at the same time creating an awareness of the need for compliance with statutory data protection requirements.

EICAR eV is the European Expert Group for IT Security (www.eicar.org), which was founded in 1991 and is recognized as a network of experts on IT security. The RFID Task Force has been part of EICAR since 2004, and the authors of this book are members of this task force. Frank Gillert is one of the co-authors of the RFID guideline, which was generated in cooperation with the German Federal Commissioner for Data Protection (BfD), BSI and GS1 Germany GmbH.

Among other things, the guideline can be used as an annex to general terms and conditions of business or project contracts related to the use of RFID technologies. Persons in decision-making positions can then determine which data protection criteria are applicable in a particular situation and what must be done to comply with them.

The data protection requirements relevant to the use of RFID tags must generally be determined based on the specific application. For this reason, three basic RFID use scenarios are described in the guideline:

1. RFID tags that are removed when the consumer acquires the goods and are thus not linked to customer data, such as RFID tags attached to garments for theft prevention.

2. Reading product data from RFID tags, such as at a point of sale (POS) for payment purposes, with RFID tags attached to retail items. In this case the tag data can be associated with customer data via a credit card or customer card, among other possibilities.

3. Tags used to generate use profiles, such as tags in customer cards.

The relevant data protection aspects are described in sections 8.3–8.5 for each of these scenarios individually.

Naturally, the authors cannot assume any liability for the correctness of these descriptions. Advice from legal experts should be obtained for assessing specific situations. The EICAR association is ready and willing to provide assistance in this regard.

8.2 General Data Protection Aspects

8.2.1 Applicability of Data Protection Legislation

Data protection legislation can be applicable if personal data is stored in RFID tags or nonpersonal data in a RFID chip can be associated with an identified or identifiable natural person. Accordingly, using RFID tags in a consumer supply chain up to the shop level does not fall within the scope of applicability of data protection legislation if the data in the system cannot be associated with individual persons.

8.2.2 Data Protection Legislation: Prohibition with Reservation for Licensing

If we examine how personal data is defined in data protection legislation, we see that personal data is defined in section 3 subsection 1 of the German Federal Data Protection Act (BDSG) as information about the personal or material circumstances of an identified or identifiable natural person. The 'prohibition principle' is applicable in data protection law. It states that collecting and processing personal data is fundamentally prohibited and only allowed by exception with the consent of the data subject or with legal authorization.

8.2.3 Principle of Data Avoidance and Data Economy (Section 3a of the BDSG)

The following applies to users that collect personal data: according to the principle of data avoidance and data economy (section 3a of the BDSG), the data controller must first consider alternatives that achieve the same objective without collection of personal data. If personal data is necessary, they must be collected openly and transparently, which means in a manner that is recognizable and comprehensible to the data subjects.

8.2.4 Data Processing in the Nonpublic Sector: Sections 27ff of the BDSG – What Is Allowed?

Data processing is admissible in the nonpublic sector without the consent of the data subject if it complies with the provisions of sections 27ff of the BDSG. Section 28 of the BDSG is particularly relevant here. It defines criteria for the admissibility of data processing for internal use. Section 28 subsection 1 no. 1 of the BDSG states that data can be used in accordance with the purposes of a contractual relationship. In addition, section 28 subsection 1 no. 2 allows use of data by way of weighing opposing interests: use of personal data is admissible without consent 'if necessary to safeguard the justified interests of the data controller and there is no reason to assume that the data subject has an overriding legitimate interest in his data being excluded from processing or use'.

8.3 Data Protection Legislation Aspects of Scenario 1

In scenario 1 as described in Section 8.1, RFID tags are used for purposes such as theft prevention. These tags are removed at department store POS locations and reused.

8.3.1 Legal Assessment

The BDSG is not applicable. A typical example of this use scenario is special RFID tags used for theft prevention in department stores. These tags are removed at the POS and reused.

8.3.2 Organization

From an organizational perspective, it is nevertheless recommended that the enterprise advise customers on the business premises that RFID tags are used for goods identification. The enterprise should provide suitable information about the technology, the exact locations where RFID tags are affixed, and the possibility of removing tags without destroying the products. This ensures proper transparency with regard to the use of RFID tags. Nevertheless, there is no explicit statutory obligation.

8.4 Data Protection Legislation Aspects of Scenario 2

Merchants wish to use RFID tags for many different purposes, such as supporting inventory processes, improving goods availability or simplifying POS processes. This corresponds to scenario 2 as described in Section 8.1. The tags used for such purposes can be disabled, for example by suppressing the readout function, or they can be removed easily.

8.4.1 Admissibility in Terms of Data Protection Legislation Pursuant to Section 28 Subsection 1 No. 1 of the BDSG

No personal data is stored directly in the RFID tags in this scenario. At the POS, product data is read out (analogous to the use of barcodes) and the relevant product data, such as the price and description of the item, are retrieved from a merchandise information system. This use is intended to reduce processing time at the point of sale. There are thus no noteworthy aspects in the case of cash payment.

However, reference to a natural person can be possible if payment is made using a customer card, bank card or credit card. The data can only be associated with a person if payment is made using a card of this sort. This situation thus falls within the scope of data protection legislation, which makes a legal basis for processing personal data necessary. This basis is provided by section 28 subsection 1 of the BDSG.

Section 28 subsection 1 clause 1 no. 1 of the BDSG permits processing of personal data if this is necessary to fulfil a contract. This provision forms the basis for data processing for fulfilment of a contract or a quasicontractual fiduciary relationship.

The collected customer data is necessary for fulfilment of a contract if payment is made using a customer card, bank card or credit card (section 28 subsection 1 clause 1 no. 1 of the BDSG), and they can be for this purpose (such as payment and/or processing a discount with a customer card) without the consent of the data subject. Nevertheless, the personal data can only be used for the specific purpose for which they were originally collected and can only be stored for as long as is necessary for this purpose (principle of limitation of use to specific purposes).

In addition, the data controller must comply with the associated data and information obligations anchored in the BDSG. Nevertheless, the consent of the customer is not necessary. Use of RFID tags in connection with bank cards, customer cards and/or credit cards is thus not objectionable with regard to data protection legislation if and as long as the information obligations anchored in the BDSG, as described below, are fulfilled and the data controller complies with all other obligations arising from data protection legislation. If these conditions are fulfilled, the process does not differ legally from current POS processes.

8.4.2   Information Obligations

Information obligations according to section 4 subsection 3 of the BDSG

Pursuant to this provision, the data controller must inform the data subject at the time of collection of the data as to:

- the identity of the body that processes the data;

- the purpose of collection, processing and use of the data;

- the categories of recipients, to the extent that the data subject has no reason to expect that the data will be transferred to them.

Information obligations according to section 6c of the BDSG

The provisions of section 6c of the BDSG are also applicable if RFID smart cards are used. They specify wide-ranging information obligations with regard to 'mobile storage and processing media for personal data'. According to section 3 subsection 10 of the BDSG, this means storage media:

- that are issued to the data subject;

- on which personal data can not only be stored but also processed automatically by the issuing body or another body;

- which allow the data subject to influence this processing only by using the medium.

This thus encompasses media on which personal data can not only be stored but also processed automatically by the issuing body or another body, or in other words, media equipped with a processor chip. However, this provision is only relevant if personal data is stored in the actual RFID tag, which is currently not the case in the trade sector.

According to section 6c subsection 1 of the BDSG, both the issuing body and the body that applies a procedure for processing on the medium must inform the data subject of their identity and address (no. 1), the mode of operation of the medium (no. 2), the data subject's right to receive information about the data and correct the data (no. 3), and the measures to be undertaken in case of loss or destruction of the medium.

In general, it must be pointed out that the information obligation must not only fulfil the provisions of section 4 subsection 3 of the BDSG and section 6c as relevant, but that it is further advisable to call attention to the exact location where the RFID tag is affixed and the possibility of removing the tag without destroying the product.

8.4.3 Technical and Organizational Measures (Section 9 of the BDSG)

According to section 9 of the BDSG, the processing body must take all technical and organizational measures necessary to ensure that personal data is processed in accordance with the provisions of data protection legislation. The effort involved in taking these measures must be reasonable relative to the intended purpose, taking into account the current state of technology. Data protection can be considered to be effective if the totality of the measures undertaken provides adequate protection against misuse of the data. The objective of protection is to ensure the confidentiality, availability, integrity and authenticity of the data.

8.4.4   Rights of the Data Subject

Notification obligation according to section 33 of the BDSG

This section provides that when the data is first stored, the data subject is to be notified of the fact of storage of the data, the type of data, the purposes of collecting, processing or using the data, and the identity of the data controller. According to section 33 subsection 1 of the BDSG, if personal data is stored in the course of business without the knowledge of the data subject for the purpose of transfer, the data subject shall be notified of the initial transfer and the type of data transferred.

From an organizational perspective, notice of the use of RFID at the entrance to a shop is thus legally required according to section 33 of the BDSG. This can most reasonably be done by means of a sign at the entrance to the shop. An application can only be assumed to be legally compliant if this general obligation to provide information with regard to the use of RFID tags in the shop is fulfilled.

Obligation to provide of information in accordance with section 34 of the BDSG

Under the provisions of section 34 of the BDSG, the data subject is also entitled to request information from the data controller. The information to be provided consists of the stored data concerning the data subject and the origin of this data, the recipients or categories of recipients to which the data is transferred, and the purpose of storage.

Disabling of RFID tags

An obligation to erase the data exists in the first instance if collection or storage of the data is inadmissible (section 35 subsection 2 no. 1 of the BDSG). In addition, erasure is required if storage is no longer necessary to fulfil the intended purpose (section 35 subsection 2 no. 3 of the BDSG). This means that an enterprise must erase data collected using RFID tags and stored in its IT system when the data is no longer necessary for fulfilling the purpose for which the data was collected. The data stored in the actual RFID tags must also be erased if the data has a personal nature. Erasing the data in the RFID tags is not at the discretion of the enterprise. However, this is presently not relevant in actual practice, since in most cases RFID tags that store personal data is not yet in use.

8.4.5   Penalties

Section 7 of the BDSG provides for compensation in case of violation of the provisions of data protection legislation. Furthermore, violations of the BDSG are classified as administrative offences (section 43 of the

BDSG) that may also be punishable under the provisions of criminal law (section 44 of the BDSG). Applications can only be regarded as legally compliant if the enterprise complies with the requirements voluntarily and on its own initiative, in particular with respect to the above-mentioned information and notification obligations regarding the use of RFID tags in shops and the possibility of disabling RFID tags after use.

8.5 Data Protection Legislation Aspects of Scenario 3

RFID tags in scenario 3 as described in Section 8.1, which is not yet relevant in practice, cannot be disabled and are used to generate personal use profiles. As the current trend in using customer relation management systems is to focus on target groups that are as homogeneous as possible, using RFID tags for this purpose is a natural idea. Consequently, criteria related to data protection should be assessed as early as the first consideration of this possibility in order to avoid creating business processes that will ultimately fail due to violation of data protection regulations.

8.5.1 Legal Assessment: Admissible Only with the Consent of the Customer

The first criterion that must be examined is the admissibility clause of section 28 of the BDSG. If the provisions stated there are fulfilled, the scenario can be regarded as legitimate. The provisions of section 28 subsection 1 clause 1 no. 1 of the BDSG (contractual purpose) are normally not fulfilled. In the case in question, collecting RFID data is evidently not necessary for fulfilling contractual obligations or exercising contractual rights. The admissibility condition of section 28 subsection 1 clause 1 no. 3 of the BDSG (data from generally accessible sources) is equally irrelevant.

However, section 28 subsection 1 no. 2 of the BDSG might be pertinent. It states that the use of personal data is admissible if this is necessary to safeguard the justified interests of the data controller and there is no reason to assume that the data subject has an overriding legitimate interest in his or her data being excluded from processing or use.

A justified interest in the sense of this provision is a 'legitimate interest, which may also be of a commercial nature, consistent with a reasonable estimation of the circumstances'. The extent to which the legitimate interest of the data subject overrides the justified interest of the data controller can be determined by weighing the respective interests.

Given the nature of the collected data and its significance on the one hand, and the interest of the data controller on the other hand, the provisions of section 28 subsection 1 no. 2 of the BDSG are normally not fulfilled when RFID is used in this particular situation. First, it is not clear why collecting information about the customer's movements in a shop is necessary in order to safeguard the justified interest of the business. Even if a justified interest of the department store is assumed to exist, this data is not necessary for safeguarding a commercial interest of the shop. Furthermore, in such situations the customer normally has an overriding interest in preventing the generation of personal use profiles. As a result, it can be concluded that neither section 28 subsection 1 no. 2 of the BDSG nor the other legal admissibility conditions of the BDSG are pertinent.

Intermediate conclusion. With this scenario, no admissibility condition satisfying the provisions of section 28 of the BDSG is present. Use of RFID tags in this situation is thus only admissible with the consent of the customer. Without the effective consent of the data subject, tracking the movements of the customer in the shop and generating use profiles constitute grave violations of the right of individual persons to determine how their personal data is used.

8.5.2   Requirement of Effective Consent

The BDSG is based on the concept of informed consent: consent must be based on a free decision of the data subject (section 4a subsection 1 clause 1 of the BDSG). The data subject must be aware of the foreseen processing in order to make a sufficiently definitive statement. The BDSG explicitly stipulates that the data subject must be informed in advance of the consequences of his or her consent (section 4a subsection 1 clause 2 of the BDSG).

Even in the case of data processing based on consent, the data controller must inform the data subject of its identity, the purpose of processing the data, and the categories of recipients (section 4 subsection 3 of the BDSG), and it must fulfil the information obligations stated in section 6c of the BDSG as appropriate. However, the data subject is usually provided with this information when consent is obtained, so separate provision of this information pursuant to section 4 subsection 3 or section 6c of the BDSG is fundamentally unnecessary. In addition, the general obligation to provide notification regarding the use of RFID tags in a shop and the obligation to provide information about the exact places where the tags are affixed and the possibility of removing the tags must be respected.

In practice, the consent of the customer can be obtained in the process of issuing a customer card. If the customer applies for a customer card, he or she can be given an opportunity to explicitly consent in writing to the use of RFID tags for the purpose of generating a use profile in the shop.

Naturally, preformulated consent statements must comply with the stringent demands of legal precedents regarding personal data processing consent clauses incorporated in general terms and conditions of business (T&Cs). Concealed inclusion of such statements in the 'fine print' is not admissible because in such cases the customer is not aware that he or she has made a declaration of consent.

Particular emphasis must therefore be given to consent by the customer. In other words, it must be presented in a clearly visible location and distinguished typographically from the rest of the text. The clause must be formulated clearly and understandably and must inform the customer of the use of RFID tags, provide a detailed description of the purpose of processing of the data, and advise the customer of his or her rights under the terms of data protection legislation. In addition, it should state when the data will be erased or the tags will be disabled after the payment transaction. In particular, the customer must be informed of the possibility of disabling the tags after use.

A customer who signs a clause of this sort is fully aware of the significance of his or her consent. The customer knows which data can be retrieved and purpose of this. The entire process is thus transparent. However, it is also advisable to provide notification of 'concealed' tags affixed to shelves, products and the like in order to comply with the full scope of the transparency principle and the supervision obligation. This corresponds to the obligation to provide information about the exact locations where tags are affixed and the possibility of removing them.

In addition, customers should be informed of their right to withdraw their consent. When a customer consents to the use of RFID tags for the purpose of generating a use profile in a shop, this action is subject to the provisions of data protection legislation. A single consent given by a customer when applying for a customer card does not constitute grounds for continuing to generate an unlimited number of customer profiles over many years. Consequently, customers should be notified at regular intervals that they have given such consent and/or requested to confirm continued extension of their consent.

In addition, the information obligations, notification right, information right and erasure right of the data subject described for scenario 2 are applicable here as well. A RFID application can only be regarded as legally compliant if these information and notification obligations and the possibility of disabling RFID tags are fulfilled by the enterprise voluntarily and on its own initiative.

The data controller must take all technical and organizational measures in accordance with section 9 of the BDSG necessary to ensure compliance with the level of data protection stipulated by law.

When RFID tags are used in the employee domain (access cards), attention must be given to compliance with statutory provisions regarding protection of employee data.

8.5.3   Penalties

The third scenario would be possible with the consent of the customer. In the absence of a suitable declaration of consent, this process would represent unauthorized collection or processing of personal data. This can be treated as an administrative offence (section 43 of the BDSG) or a criminal offence (section 44 of the BDSG). The possibility that the data subject may claim compensation can also not be excluded (section 7 of the BDSG).

9

RFID Legislation in a Global Perspective[1]

9.1 RFID Scenarios

Radio frequency identification (RFID) technology provides opportunities to improve products and services, make them safer and cheaper, and even protect people and animals from themselves and others [Floer04].[2] RFID presents global and ubiquitous challenges to legal systems. The challenges are global because the products in which RFID technology is integrated and RFID technology itself are intended to be marketed worldwide and will be sold worldwide. RFID technology is ubiquitous because its scope extends from integration into products to implantation in people and animals. The magnitude of potential applications for RFID technology forces us to distinguish several different scenarios from a legal perspective. Accordingly, we must distinguish whether RFID technology is used for:

- identification and monitoring of products (electronic product code (EPC) scenario);

- identification and monitoring of animals (real-time authentication and monitoring of animals (RTAMA) scenario);

- identification and monitoring of people (real-time authentication and monitoring of persons (RTAMP) scenario).

[1] Contributed by Professor Viola Schmid LLM (Harvard). Professor Schmid assumed the Chair of Public (International) Law in the Faculty of Law and Economics of the Darmstadt University of Technology (Darmstadt, Germany) in 2002. Her research areas are cyberlaw, e-justice and freedom of speech. Contact: schmid@jus.tu-darmstadt.de.

[2] Floerkemeier et al. identify more than 15 types of purpose declarations for RFID reader queries: access control, anticounterfeiting, antitheft, asset management, contact, current, development, emergency services, inventory, legal, payment, profiling, repairs and returns, and other.

An example of the RTAMP scenario is the Legoland KidSpotter in Billund, Denmark [STOA07]:

> At the entrance to the park, parents can rent a wristband containing an active RFID device for their children for EUR 3 per day. Around 40 to 50 RFID readers have been placed throughout the 150,000 square-metre park. If parents lose sight of their child, they can send a text message to the KidSpotter system. They will receive a return message with the name of the park area where the child is located and the map coordinates of their child's position in the park, with an accuracy of 3 metres. This security function is the main reason for parents renting the wristband to combat the problem that about 1600 children get lost in the park each year. Identity management in this case involves a combination of personal identity, place and phone number.

Other examples of RTAMP are tracking school children or mine workers [TStar07] whose locations can be determined using RFID. In this context, RFID technology has the potential to save lives. The RTAMP scenario is an intriguing one from the perspective of information privacy law because it poses the question of whether it is permissible to tag persons, possibly even against their will. In this regard, two states in the USA – Wisconsin and North Dakota – passed laws in 2006 and 2007 that prohibit forcing an employee to allow implantation of an RFID device.

The RTAMA scenario – such as embedding chips in deer (Texas, USA) or dogs (Hamburg, Germany) – is also a subject of public debate because people are afraid that the implants could cause cancer [Hei07].

Despite the many ethical, economical and legal aspects of these two scenarios, the following discussion concentrates on the EPC scenario, which means fitting products with RFID devices. The EPC scenario makes it possible to read out passive tags and combine their content (e.g. 'this is margarine package no. xxx') with an associated manufacturer database that provides further information about the product such as when and where it was produced (e.g. Plant X on 10 January 2008) [Lan08]. Tag readability, or in other words content privacy, is not the only issue that forms a challenge to RFID law. It is also possible to place RFID readers in various locations (such as in a supermarket) to determine where a particular product is located at a particular time (which affects locational privacy). Even with this relatively simple scenario, the statements of a passionate anti-RFID activist on the one hand and a renowned scientist on the other hand give an indication of the crisis into which modern law can be plunged by RFID technology. Katherine Albrecht of Consumers Against Shopping Privacy Invasion and Numbering (CASPIAN) asserts that 'the risks RFID poses to the social world are comparable to the risks nuclear weapons pose to the physical world. In the same way that bombs destroy objects, RFID could decimate privacy' [Alb07]. A representative of the scientific establishment, Marc Langheinrich of the Swiss Federal Institute of Technology, makes the following provocative statement: 'Perhaps the question in the future will not have to be "Do we still have privacy?", but instead "Do we still want to have privacy?" ' [Lan08]. These two views of RFID technology could not be more diametrically opposed: Katherine Albrecht wants to protect us against RFID, while Marc Langheinrich wants to use RFID in smart products, environments, trolleys, cars and phones. Both views demand an analysis of how privacy is affected by the use of RFID technology.

9.2 Fair Information Practices for Personal Data: Traditional Law

From a global perspective, the basis for dealing with RFID technology is formed by what are called 'fair information practices', which are expressed in the following documents, among others:

- United Nations *Guidelines Concerning Computerized Personal Data Files* [UN90].

- Organization for Economic Cooperation and Development *Guidelines Governing the Protection of Privacy and Transborder Flows of Personal Data* [OECD80].

- Council of Europe *Convention for the Protection of Individuals with Regard to Automatic Processing of Personal Data* [CE81].

These guidelines are inspired by various sources, such as the US Fair Information Principles, which were developed by the US Department of Health, Education and Welfare in 1973 [USFIP73]. In Europe, these partially legally nonbinding guidelines are complemented by two binding directives dealing with protection of personal data [EU95/02], which have been implemented in the form of data protection and IT security acts in the individual member states. The core of this legislation reflects the eight principles of the OECD guidelines:

1. **Collection Limitation Principle.** There should be limits to the collection of personal data, and any such data should be obtained by lawful and fair means and, where appropriate, with the knowledge or consent of the data subject.

2. **Data Quality Principle.** Personal data should be relevant to the purposes for which they are to be used, and to the extent necessary for these purposes, should be accurate, complete and kept up to date.

3. **Purpose Specification Principle.** The purposes for which personal data is collected should be specified not later than at the time of data

collection, and subsequent use should be limited to fulfilment of these purposes or such others as are not incompatible with those purposes and as are specified on each occasion of change of purpose.

4. **Use Limitation Principle.** Personal data should not be disclosed, made available or otherwise used for purposes other than those specified... except:

   (a) with the consent of the data subject, or

   (b) by the authority of law.

5. **Security Safeguards Principle.** Personal data should be protected by reasonable security safeguards against such risks as loss or unauthorised access, destruction, use, modification or disclosure of data.

6. **Openness Principle.** There should be a general policy of openness about developments, practices and policies with respect to personal data. Means should be readily available for establishing the existence and nature of personal data and the main purpose of their use, as well as the identity and usual residence of the data controller.

7. **Individual Participation Principle.** An individual should have the right:

   (a) to obtain from a data controller, or otherwise, confirmation of whether the data controller has data relating to him;

   (b) to have communicated to him data relating to him (within reasonable time; at a charge, if any, that is not excessive; in a reasonable manner; and in a form that is ready intelligible to him);

   (c) to be given reasons if a request made under subparagraphs (a) and (b) is denied, and to be able to challenge such denial; and

   (d) to challenge data relating to him and, if the challenge is successful, to have the data erased, rectified, completed, or amended.

8. **Accountability Principle.** A data controller should be accountable for complying with measures which give effect to the principles stated above.

In summary, it can be stated that whenever personal data is involved, the fact that this data is collected and used must be communicated and effective measures must be taken to protect them against skimming and eavesdropping. Traditional law thus already contains the appeal that surreptitious reading of RFID data is not allowed and protection must be provided against unauthorized access to and/or modification of the data.

9.3   Fair Information Practices for RFID Data

9.3.1   EPC Scenario Not Involving Personal Data

The concept of personal data is a core aspect of information privacy law. Traditional privacy law only provides protection when such personal data is collected or read. However, in the EPC scenario it is questionable whether personal data is collected when a consumer carries a tagged item that only contains a number and that can be read with a suitable reader to obtain information about the manufacturer, plant and production date. Linking a person to one or many such items of information, regardless of how many tagged garments the person may be wearing or how many products are in the shopping cart, does not automatically result in personal data if the person concerned is not specifically identified. This means that traditional law cannot provide a solution for the EPC scenario. As a result, the confrontation between RFID technology and privacy leads to two different responses:

- technological solutions (privacy-enhancing technologies);
- the question as to whether new RFID law is needed.

9.3.2   Privacy-Enhancing Technologies (PETs)

The legal analysis presented here mentions only a few of the contemplated technical strategies. Further details are described the relevant computer science literature [Lan05; Lan08].

RFID tags labelled for recognition and removal

The first solution is to mark the RFID tag with a label, which also makes the tag removable. This solution corresponds to the legally nonbinding EPC guidelines of the EPCglobal industry association [EPC2005]. The idea of only using RFID tags on packaging and making them removable also corresponds to the wishes of 471 out of 2014 respondents to an online survey conducted by the European Commission in 2006 [EUCo07a].

Anonymization by means of a kill command

Another possibility is to equip the tag with a kill function. Of course, this kill function must be password-protected so the tags cannot be erased without authorization. Practical problems arise with the kill function if it is not possible to prove to customers that the tags can in fact no longer be read, so that tracking and RFID searching are no longer possible [Lan05].

Pseudonymization by means of cryptographic methods

In this case hash values are generated in order to control the readability and trackability of the RFID tags.

Blocker tags

If a person carries a blocker tag, unauthorized reading of RFID tags is no longer possible. However, the effectiveness of this approach is disputed [Foe07].

Metal envelopes

RFID technology requires interference-free frequencies. Fitting wallets with metal foil linings could prevent reading of bank notes or identification documents containing RFID tags.

Prospects

In summary, it can be stated that technological options are available but additional investigation of their economic, technical and legal practicality is necessary. For example, blocker tags could interfere with desirable RFID searches, with the result that future RFID law might prohibit the use of blocker tags. Beyond the demand for RFID privacy technology, there is thus a demand for new law.

9.4   RFID Law

9.4.1   Some Insights into Legal Research

EPC scenarios demand a global perspective, but this global approach is foreign to the nature of law, which is the product of sovereign states and supranational associations (such as the European Union). This foreignness is starting to disappear with the publication of legal works on the Internet, and at least in the case of the European legal realm (http://eur-lex.europa.eu/de/index.htm), the legal realm of the USA at the federal (http://thomas.loc.gov/) and state (http://www.llrx.com/columns/roundup23.htm#state) levels, and the Australian legal realm (http://www.austlii.edu.au/), generally accessible databases facilitate research on legislative policy and legislation. Utilization of these Internet resources will be of decisive importance for researching RFID law, which is still in the formative stages. However, it must be admitted at this point that the research described below is strictly Internet-based and is thus not based on legally authentic sources. Nevertheless, these sources are recommended here because they should make it easier for readers of this book to investigate the state of legislation in the future by means of their own research.

9.4.2   The USA as a Pioneer in RFID Legislation?

As far as can be seen, the worldwide pioneer in RFID legislation is the USA at both the federal and state levels. Two states (Wisconsin and North Dakota) prohibit implantation of RFID devices in persons against their will. On the other hand, RFID is recommended at the federal level for tracking and tracing of prescription drugs [HR3580]:

Food and Drug Administration Amendments Act of 2007 [HR3580]

(a) **In General.** The Secretary shall develop standards and identify and validate effective technologies for the purpose of securing the drug supply chain against counterfeit, diverted, subpotent, substandard, adulterated, misbranded, or expired drugs.

(b) **Standards Development**

1. **In General.** The Secretary shall, in consultation with the agencies... manufacturers, distributors, pharmacies and other supply chain stakeholders, prioritize and develop standards for the identification, validation, authentication, and tracking and tracing of prescription drugs.

2. **Standardized Numerical Identifier.** Not later than 30 months after the date of the enactment of the Food and Drug Administration Amendments Act of 2007, the Secretary shall develop a standardized numerical identifier (which, to the extent practicable, shall be harmonized with international consensus standards for such an identifier) to be applied to a prescription drug at the point of manufacturing and repackaging (in which case the numerical identifier shall be linked to the numerical identifier applied at the point of manufacturing) at the package or pallet level, sufficient to facilitate the identification, validation, authentication, and tracking and tracing of the prescription drug.

3. **Promising Technologies.** The standards developed under this subsection shall address promising technologies, which may include:

   (A) radio frequency identification technology;

   (B) nanotechnology;

   (C) encryption technologies; and

   (D) other track-and-trace or authentication technologies.

RFID is also presently a subject of legislative attention in 18 states of the USA [Schm08]. A prominent example of this is the 'right to know' legislation controversy, which was occupying the attention of the legislative community of New York in 2007.

9.4.3 'Right to Know' Legislation in New York (2007)

This draft legislation [NY07] requires every 'retail mercantile establishment that sells or offers for sale merchandise containing radio frequency tags' to observe the following five commandments:

1. **Notice**... such notice shall disclose:

   – that the establishment offers items with RFID tags, and

     – that New York State law requires the establishment to remove or disable all radio frequency identification tags before tagged items leave the establishment, and

     – that the establishment is required to provide consumers on request, with personal information gathered within the establishment through the RFID tags used in the establishment.

2. **Labelling**... no retail mercantile establishment shall sell... any item... that contains... an RFID tag unless such item... is labelled with a notice stating that such item... contains or bears a RFID tag, and that the RFID tag can transmit unique identification information to an independent reader before and after the purchase. Such label shall be posted on the item or package in a conspicuous type, size and location and in print that contrasts with the background against which it appears.

3. **Information** Upon written request of a consumer, a retail mercantile establishment that has gathered personal information through radio frequency identification tags shall release to the requester all of the stored personal information pertaining to the requester. Every retail mercantile establishment shall make available to consumers a form for such requests.

4. **Removal or deactivation** Every retail mercantile establishment that offers items or packages that contain or bear radio frequency identification tags shall remove or deactivate all tags at the point of sale... In addition: all costs of whatsoever name or nature for the removal or deactivation of a radio frequency identification tag shall be borne by the retail mercantile establishment... A retail mercantile establishment shall not coerce consumers into keeping radio frequency identification tags on items or packages by requiring items or packages to be exchanged, returned, repaired or serviced to contain or bear active tags, and a radio frequency identification tag, once removed or deactivated, shall not be reactivated without express consent of the consumer associated with the tagged item.

5. **No aggregation of personal information and RFID tag information** No retail mercantile establishment shall combine or link a consumer's personal information with information gathered by, or contained within, a radio frequency identification tag... No retail mercantile establishment shall, directly or through an affiliate, disclose to a nonaffiliated third party a consumer's personal information associated with information gathered by, or contained within, a radio frequency identification tag.

The fifth stipulation in particular – the prohibition on customer profiling by combining EPCs with personal data – arises from the fears of RFID opponents. They want RFID devices to become part of the 'internet of things' but not of an 'internet of persons'. Just as it should be possible to represent things digitally and in real time (real-time enterprise and real-world awareness), it should not be possible to use the same technology to digitize people at all places and at all times and thus turn them into 'real-time persons'. It can also be seen from the New York initiative that it includes a stipulation for consumer information and deactivation or removal, even for the nonpersonal data of an EPC scenario. This is a new form of RFID law that leaps over the personal data hurdles of traditional law. Here it must be said that it is not at all certain that the State of New York will actually enact such a 'right to know' act. A similar legislative initiative in California failed in 2005. Nevertheless, a global perspective reveals that legislators on other continents are also wrestling with RFID 'right to know' law similar to the New York model. For example, it is no coincidence that privacy commissioners worldwide have demanded legislative action [WPC03].

9.4.4 'Right to Know' Legislative Policy on Other Continents

Internet research yielded results for Europe, Germany, Japan, Korea, Australia and South Africa.

RFID legislation and the European Community

The European Community has two directives that specify how personal data (including locational privacy) must be handled. There is thus already data protection law that applies to RFID tags that contain personal data. In addition, the Working Document (2005) of the Article 29 Group of data protection commissioners, which is based on Directive 95/46/EC, essentially stipulates compliance with the same commandments as in New York [Art05]. The European Commission is aware that RFID is an explosive issue with regard to privacy and has announced that guidelines for the public and private sectors will be issued in late 2007. It has also said that it is willing to review the need for amendments to the previously mentioned directives [EUCo07b]. Overall, the path being taken by the European Community is presently characterized by an awareness of the problem, but – unlike the USA – an absence of specific legislative initiatives as yet.

RFID legislation in Germany

Hessen, which is one of the 16 German federal states, is an international pioneer in data protection. A data protection act has been in force in Hessen since 1970. Nevertheless, up to now (2007) the German federal legislative body (the Bundestag) has only resolved to ask the government to examine the issue. This hesitation on the part of legislators contrasts with the resolutions of the federal and state data protection commissioners [Kon06] and the highest supervisory authorities for data protection in the private sector [Auf06], which demand action such as in New York.

RFID legislation in Japan

Japan has issued an 'Act on the Protection of Personal Information', but it is a form of traditional law that only deals with RFID devices containing personal data. In 2004, the Ministry of Internal Affairs and Communications (MIC) and the Ministry of Economy, Trade and Industry (METI) issued a document titled 'Guidelines for Privacy Protection with Regard to RFID Tags' [Jap04]. These guidelines are based on the recommendations of a working group for improving product traceability. In terms of their content, these legally nonbinding guidelines correspond in part to the New York model.

RFID legislation in the Republic of Korea

Korea also does not yet have any RFID legislation, but only a sort of soft law in the form of an 'RFID Privacy Protection Guideline' (revised September 2007) [Ko07]. It acts as a self-regulatory guideline for RFID service providers and follows the New York model in key points with its information and deactivation obligations.

RFID legislation in Australia

No RFID legislation in Australia is available for searching. However, the RFID commandments of the New York model are included in an issues paper generated by the Australian Law Reform Commission in 2006 [Au06].

RFID privacy concerns in South Africa

For South Africa, research turned up a government notice (26 August 2006) [SA06] of the Independent Communications Authority of South Africa, which raises the question of privacy and IT security in connection with spectrum reallocation to cater for RFID ('the issues of authenticity, integrity as well as privacy are often raised in the context of RFID'). However, no legislative efforts on the African continent could be found by searching.

9.5   RFID and Law: A Summary of the Situation in 2007

This chapter shows that RFID cannot be associated with nuclear weapons (as Katherine Albrecht suggests) or surrendering the private realm (as Marc Langheinrich suggests). In countries that provide protection for personal data, such as the 27 member states of the European Community, traditional law already provides protection against using RFID for geographic localization (locational privacy) and personal profiling. Whether new law for RFID is needed in addition to this is presently (2007) a subject of debate in many countries. The five commandments of the legislative policy of New York can also be found in the soft law of Korea and Japan and the legislative recommendations of chief privacy officers and the Australian Law Reform Commission. Further developments, especially in the USA (which is a pioneer in this regard), will show whether RFID technology will also lead to legal protection for collecting and transferring nonpersonal data such as EPCs (as in the New York model).

10

Applications

Proof that a new technology works is provided by successful use of the technology and the benefits arising from its use. In many respects, RFID technology is still in the early stages of development and successful applications are rare. Consequently, the applications described in this chapter include some examples that still have the character of pilot projects but nevertheless have major potential for widespread use.

Some typical features of the applications described in this chapter are summarized in Table 10.1. The item numbers in the table correspond to section numbers 10.1 to 10.12. This makes it easier to identify the cases that readers may be interested in and which most closely correspond to interests with regard to using RFID.

Application cases 1 (logistics and Hewlett-Packard) and 2 (mobile maintenance at Fraport) were taken from the Bitkom RFID White Paper and updated [Bitkom2005].

10.1 Logistics Processes at Hewlett-Packard

With a spectrum of more than 23 000 different products and trade relationships with more than 100 000 suppliers, Hewlett-Packard (HP) has been using RFID methods for a long time already. They help reduce warehousing (Figure 10.1) and shipping costs in the company's own supply chain and the costs of its trading and sales partners.

RFID is presently used in 30 plants and logistics centres, primarily in China, Southeast Asia, the USA and Brazil. HP originally introduced RFID to optimize its own logistics processes. When Wal-Mart demanded labelling with RFID tags, HP was one of the first suppliers to be able to meet this requirement.

HP uses RFID for purposes such as tagging at the shipping unit and pallet levels in its plant in São Paulo (Brazil) and its logistics centres in

RFID for the Optimization of Business Processes   Wolf-Ruediger Hansen and Frank Gillert
© 2008 John Wiley & Sons, Ltd

Table 10.1 Typical features of the applications described in this chapter

	Application	Technology	Scope	Sector	Status
1	Logistics processes at Hewlett-Packard	RFID	Global	Trade	Operational
2	Mobile maintenance at Fraport	RFID	Local	Maintenance	Operational
3	Car location at Dat Autohus	RFID, GPS	Local	Trade	Operational
4	Locating persons in hazard areas in the Gotthard Tunnel	RFID	Local	Personal protection	Operational
5	Electronic ticketing in public transport systems	RFID	Regional	Consumer	Operational
6	Application scenarios for NFC mobile telephones	RFID	Global	Consumer	Operational/pilot
7	Monitoring components in computer centres	RFID	Local	Asset management	Pilot
8	Aircraft seat Quick-Change at Lufthansa	RFID	Local	Aviation	Operational
9	Tracking trolleys at the Finnish Post Office	RFID	National	Logistics	Operational
10	Tracking products and containers at KPN in the Netherlands	RFID	National	Trade	Pilot
11	Tracking garments at Gardeur	RFID	International	Garment logistics	Operational
12	Agent technology in intralogistics systems	RFID	Local	Manufacturing	Pilot

Chester and Memphis (USA). The São Paulo plant produces printers and ships them to several logistics centres. Ink cartridges are manufactured, packaged and stored temporarily in Chester, and they are shipped to the Memphis logistics centre. All cases are fitted with RFID tags to improve business processes in the warehouses and shipment of high-value printers.

Because of the use of RFID, the logistics process time for despatching pallets in Memphis has been reduced from minutes to seconds because

Antenna gate

RFID antennas
(total of 4 antennas on each side of the gate)

Palette with cases fitted with RFID tags

Figure 10.1 Warehouse exit with RFID antennas at the HP plant in Chester, USA (source: HP)

it is no longer necessary to use cumbersome barcodes. In Chester, the cost of handling shipping units and pallets inside the distribution centre has been reduced considerably by eliminating the cost drivers, losses and errors associated with manual handling. Other possibilities, such as utilizing supplementary information extending beyond standard EPC data, have not yet been exploited fully.

Worldwide use of RFID in HP's supply chain is being developed further and implemented systematically in collaboration with logistics enterprises, trading partners and major customers. Step by step, other sites are joining the ranks of the 30 HP plants and logistics centres that already use RFID. Additional efficiency gains are expected to be realized from introducing second-generation (EPC Gen 2) RFID transponders.

10.2   Mobile Maintenance at Fraport

Fraport AG in Frankfurt, Germany, with worldwide sales of €1.8 billion, is a leader in international airport operation activities. It is also the owner and operator of Frankfurt Airport. Its business areas – ground handling, traffic and terminal management, communication services, and real estate and facility management – span the full spectrum of products and services in the airport business.

A mobile RFID maintenance scenario has been implemented here under the leadership of SAP Consulting. By linking SAP Mobile Asset Management software with portable hand-held computers and RFID technology, Fraport created an innovative system that provides legally watertight organization of maintenance activities for a wide variety of technical components at the airport site, such as fire dampers.

The mobile RFID maintenance scenario increases security at Frankfurt Airport. It replaces paperwork, communicates detailed real-time data, and closes information gaps between back-end systems and on-site operations. The RFID tags in this system are attached to the fire dampers of the buildings and are thus immobile. The maintenance technicians must approach them with their RFID readers in order to read them.

10.2.1   Optimized Paperless Maintenance Processes

As part of its real estate and facility management activities, Fraport AG manages around 420 buildings and facilities at Frankfurt Airport. One of its core tasks is periodic service and inspection of technical components with legally specified maintenance requirements. As the operator of Frankfurt Airport, Fraport is obliged to document these activities. The company uses SAP Plant Maintenance to manage its maintenance activities. This application has been extended to include a mobile system called Mobile Asset Management. Here RFID transponders operating at 13.56 MHz and supplied by Microsensys GmbH, Erfurt, Germany, and robust hand-held computers interact seamlessly to form an effective total system (see Figure 10.2, which originates from a book by Claus Heinrich [Hein2005]).

Instead of paper-based work instructions, maintenance technicians now use hand-held computers as front-end units that can read and write RFID transponders and communicate with the Mobile Asset Management application. The mobile terminals receive current maintenance orders from Fraport's central SAP system. After user identification and login, a guided step-by-step dialogue is initiated between the Mobile Asset Management application, the technician, and the RFID tags fitted to the airport facilities and systems. This procedure ensures that maintenance is performed using precisely defined process steps. Any faults that may be present are logged using comprehensive damage codes.

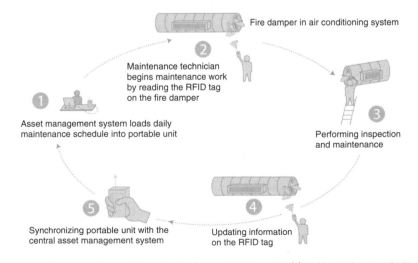

Figure 10.2   Paperless, RFID-optimized maintenance at Frankfurt Airport (source: SAP)

The history of every maintenance activity and local status are thus documented fully and recorded in electronic form. After completion of the maintenance work, the mobile terminal is read out and the data is transferred electronically to the back-end system. Overall, Mobile Asset Management has significantly accelerated maintenance processes, distinctly enhanced data quality and security, and created new possibilities for reporting malfunctions and product service life information.

With this SAP solution, Fraport AG has obtained a future-proof, modern and mobile maintenance system that can be adapted to any desired task within the service and maintenance organization. The SAP Mobile Asset Management system has been operational since May 2005, including RFID functionality as a standard feature.

10.2.2   Six-Figure Savings Realized

Fraport has acquired an ideal electronic maintenance process that enables it to guarantee correct performance of facility maintenance. Because of the use of innovative technologies, maintenance activities can be carried out faster than before. Each step of the process can now be executed precisely with the push of a button. This creates the best possible transparency for providing proof of proper maintenance to auditing bodies.

Fraport has attached RFID tags to 22 000 fire dampers at the airport. Previously, 88 000 maintenance work orders were generated each year for the dampers. Thanks to the use of RFID, Fraport AG saves around €450 000 per year on documentation, as reported by Dr Roland Krieg, CIO of Fraport AG, in an interview with *Computerwoche* [Fraport2005]. According to him, this is offset by a nonrecurring cost of less than €100 000 for the RFID tags and readers. That adds up to a very satisfactory return on investment (ROI) in less than one year.

10.3   Locating Cars at Dat Autohus

Dat Autohus AG is the largest used-car dealer in Europe. It has an inventory of more than 3000 cars for sale at two display lots north of Hamburg (Germany) in Bockel (55 000 m^2) and Sittensen (10 000 m^2). Each day 80 vehicles on average are accepted and prepared for resale. Each incoming vehicle requires more or less extensive preparation in the in-house garage. Vehicles that are ready for sale are parked on the lot (Figure 10.3) sorted by vehicle type. The next available parking space is used each time.

The sales process is largely oriented toward self-service by potential purchasers. They can learn about the available vehicles via the Internet or in the sales room and select the cars they would like to see.

Figure 10.3 A sales lot with cars located using GPS and RFID (source: Dat Autohus)

They then receive the keys to these cars and a site plan with the locations of the desired vehicles marked automatically. Previously, searching for these vehicles was often confusing, and in many cases potential purchasers simply could not find them, so they left the site disappointed and without making a purchase. One way to achieve a significant reduction in the number of disappointed shoppers is to use RFID methods to improve the chances of finding the desired vehicles.

There are several reasons why potential purchasers cannot find the desired vehicles:

- The sought vehicle is located in a group of similar vehicles and is difficult to identify by appearance.

- The position information is not sufficiently exact.

- The vehicle has been used before by another potential purchaser and parked in a different place, but this information is not available in the sales room.

10.3.1 A Solution Using GPS and RFID

To eliminate these problems, a position determination method was developed and implemented. It is based on combining GPS and RFID technologies with a software program called tagpilot from Silverstroke AG (Ettlingen, Germany). Barcodes are also used, for example to identify car keys. This approach improves three different processes:

- vehicle receiving process;

- sales process;

- back-office process.

Figure 10.4   A RFID transponder fitted between headrest supports (source: Dat Autohus)

When a vehicle is received, it is checked to determine what maintenance, service and repair work must be performed before it can be placed on the sales lot. These activities are entered into the tagpilot system. When each car is received, it is fitted with an active RFID transponder operating at 868 MHz and attached to the headrest with cable ties (Figure 10.4). The transponder also has a barcode, which is read at the same time as the transponder. This associates the barcode and transponder numbers in the IT system with the vehicle. This registration process takes place during vehicle receiving. The transponder is removed before the vehicle is handed over to the purchaser and used afterward with another vehicle.

After completion of the receiving process, the vehicle goes to the garage waiting lot. This transfer is monitored by a traffic light that turns green if the RFID transponder in the vehicle is recognized and vehicle registration is confirmed by the IT system. The planned work is performed in the garage and recorded in the IT system. After the work is completed, the vehicle is released for sale and parked at the next available position on the sales lot.

But how is the position of the vehicle determined so this information can be given to potential purchasers to help them find the vehicle? This is done by having an employee of Dat Autohus drive through the entire sales lot periodically, row by row. The employee's vehicle is fitted with a mobile RFID reader that recognizes the car transponders and determines their geographic positions using GPS. The GPS position is determined continuously during the survey round, and the RFID antenna recognizes the transponders in the vehicles parked along the rows at the same time. In this way, each vehicle is assigned a GPS position with an accuracy of 5 to 10 m, which is transmitted to the IT system via a wireless network (WLAN). As a result, the positions of all available vehicles are known after each survey trip (Figure 10.5). Position changes and the locations of newly prepared vehicles are updated on each mobile inventory cycle. The car keys are registered with barcode numbers after each car is prepared for sale.

Potential purchasers first identify cars they would like to see, either via the Internet at home or in the salesroom, and then state the numbers

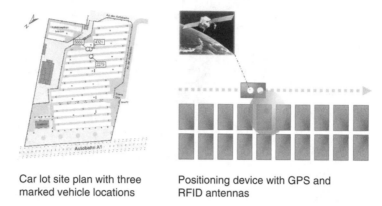

Car lot site plan with three   Positioning device with GPS and
marked vehicle locations   RFID antennas

Figure 10.5   Position determination on a car sales lot (source: Silverstroke AG)

of these cars. After the particulars of the potential purchasers have been recorded, they receive the keys of the desired vehicles and a site plan. The keys are associated with the potential purchasers by reading their barcodes again. If a vehicle whose keys have been given to a potential purchaser attempts to leave the lot, it is recognized by an RFID antenna and an alarm is triggered. Vehicles requiring off-site painting or service and vehicles that must be taken to a different site can leave the site unhindered after suitable arrangements are made in the IT system.

After viewing the desired vehicles, the potential purchaser returns the keys to the salesroom. The keys are again identified by their barcodes. If no purchase is made, the reason for this is also recorded if possible. This information is used to generate statistics that can be used to determine why certain vehicles remain unsold for unusually long times.

When a vehicle is sold, the RFID transponder is removed and read in the showroom during processing of the sales formalities. The vehicle is then deleted from the database of vehicles for sale, and the transponder is made available for reuse. In the future, the vehicle will also be removed from the website automatically.

Evaluation lists are also generated to improve insights into the sales process and customer behaviour. A daily log shows which vehicles were viewed by potential purchasers during the day. Vehicle statistics show which vehicles have been viewed previously but not yet sold. They can be used to obtain useful insights for improving the sales process.

In order to operate the system, the company purchased 3000 active transponders at approximately €50 each, two mobile GPS/RFID units, seven RFID acquisition units, 19 fixed antennas and the Silverstroke software, for a total investment of approximately €300 000 for the entire system including installation and start-up activities. The economic benefits arise from boosting vehicle turnover by improving the above-mentioned processes and the anticipated significant increase in ratio of sales to visits by potential purchasers.

10.4   Locating Persons in Hazard Areas

The TeraTron Local Positioning System (LPS) [Terat2005] is being used for the first time in the new Gotthard Tunnel in Switzerland, which is presently under construction and will be 57 km long when finished. The LPS is used to determine the exact positions of tunnel workers, with the specific aim of accurately locating workers in hazard areas. This is intended to enable precise execution of time-critical rescue efforts and avoid unnecessary efforts in areas where nobody is present. LPS achieves 100% identification. It is thus the only system in the world that fulfils the operator's requirements, which is why it was selected.

The second Gotthard tunnel presently being built is the longest railway tunnel in the world. Protection of tunnel workers (miners) has top priority. The fatality rate during construction of the first Gotthard tunnel was 10 workers per kilometre of tunnel progress, but to date only one fatality has occurred during the construction of the new tunnel. In order to further improve the safety record and allow people to be rescued quickly and reliably in case of an emergency, the operator decided to use a system with active RFID transponders to determine the locations of all workers in the tunnel. Position markers are installed at all critical points, such as stations, portals, the shaft head, the base of the shaft and in the tunnels. Workers carry active RFID transponders and are recognized automatically at these locations. Display screens in the control room allow the past and present locations of all workers to be checked at all times. In addition, the LPS can be used to ensure that only authorized persons are present in secure areas.

The same system is used by other companies, such as Kronos Titan, Leverkusen (Germany) in the chemical sector for production of hazardous materials. Here again the objective is to know exactly where people are located in case of a hazardous situation in order to undertake specific, localized rescue efforts.

System operation is marvellously simple. The system consists of four components:

1. **Position markers.** These are RFID antennas that generate a localized magnetic field with a diameter of a few metres. The frequency is 125 kHz (low frequency, LF), so it does not pose any health hazard. The fields are configured such that persons only pass through them and do not remain inside the range of the field. Each magnetic field has an ID code for unambiguous position identification and can detect the direction of travel, so it is possible to recognize whether a person enters or leaves an area.

2. **Personal RFID cards.** Standard-size smart cards or key fobs can be used for personal identification. They contain two RFID transponders. One of them is tuned to the LF frequency band of the position

markers and recognizes them when the person passes by. It then transmits a signal to the other transponder in the key fob, which operates at 868 MHz in the UHF band. The second transponder transmits current position data and a personal ID to the nearest fixed receiver via a wireless RFID link. The UHF transponder is active for only a fraction of a second, after which it returns to sleep mode.

3. **Fixed receivers.** The receivers are distributed over the area to be monitored such that it is always possible to receive the data transmitted by the UHF transponders in the personal RFID cards. They pass this data on to a program running on the server.

4. **Workstation software.** The position data is collected by a workstation and displayed in graphic or tabular form on a monitor. If an emergency occurs, the locations of persons in the hazard area are always known immediately.

The interactions between these components are shown in Figure 10.6 using a factory hall with rooms and hallways as an example. This can easily be extended to the floors of an office building or other areas. In the figure, persons are shown by heads with helmets and position markers by dots inside grey circles that represent the antenna fields. Naturally, such a system can also be used to determine the locations of other types of objects, such as portable systems or boxes of valuable goods accompanied by suitable smart cards.

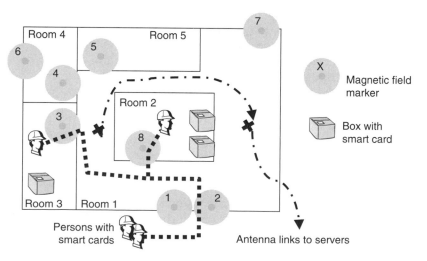

Figure 10.6   A local positioning system (LPS) for persons and objects (source: TeraTron)

10.5   Electronic Ticketing in Public Transport Systems

RFID tags create virtual identities for the objects to which they are attached, and smart cards equipped with RFID tags do the same thing for persons. They can both be identified using radio signals. RFID smart cards can be read without physical contact, which makes them different from previous types of cards read by magnetic stripe readers or via direct electrical contact with the chip in the card.

In logistics systems, it is desirable to have a reading range of several metres between antennas and objects for easy identification of cases and pallets that are pushed through antenna gates. By contrast, proximity technology is used with smart cards to restrict the reading range to less than 10 cm. This is also called 'near-field communication' (NFC), and it is based on the ISO 14 443 and ISO 18 092 standards.

People are accustomed to using magnetic stripe cards such as credit cards and smart cards such as electronic purse cards. As is well known, it takes a certain amount of time to read data from these cards in order to initiate the desired transaction. With smart cards, this can be seen during payment transactions at points of sale (POS) and at exits from multistorey car parks. Consequently, this technology is not suitable for use in public transport systems. It would severely hamper free entry and exit from public transport vehicles, and the mechanical stress of repeated reading would quickly lead to worn-out cards.

10.5.1   Pilot Project in Hanau

In 2002 the Rhein-Main Verkehrsverbund (RMV) and Hanauer Straßenbahn AG jointly started issuing contactless smart cards for use in the public transport system of Hanau, which is close to Frankfurt (Germany). In a pilot project, these smart cards were marketed under the product name 'get≫in' as media for electronic tickets. Passengers who use these smart cards must use smart card readers in the buses to register the beginning and end of their travel (Figure 10.7).

Financial settlement for the travel services used takes place afterwards (postpaid system) based on the most favourable ('best price') fare in each case. The get≫in card is a dual-interface card, which means it has a contact interface as well as a contactless interface, so customers can use simple pocket readers to check the data stored in the card. This largely avoids frequent mechanical stress. However, at first many cardholders had to learn that the cards are not disposable and that it is not advisable to trim them to fit their wallets or punch holes in them so they can be attached to a strap. Both of these actions make the cards nonfunctional and thus worthless.

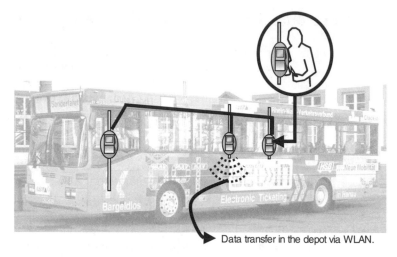

Data transfer in the depot via WLAN.

Figure 10.7 A bus in Hanau with check-in and check-out terminals for smart cards (source: RMV)

Besides smart cards, mobile telephones with NFC capability can also be used. They operate with the terminals in the buses and tramcars in exactly the same way. The advantage of this for passengers is that they do not have to carry a separate smart card.

10.5.2 Clear Business Objectives

The long-term goal of RMV is to eliminate paper tickets as much as possible. Besides faster processing of sales transactions in local public transport systems, the solution targets several other business objectives:

- Offering new fares and products.

- Optimizing marketing structures in a combined local transport authority composed of numerous local operators (local public-law corporations).

- Better analysis of revenue structure and service demand and improved planning of services.

- Improved customer relationship management.

- Realizing new utilization potentials.

There are also plans for using the smart cards for other services in the mobility, culture and recreation sphere:

- Car rental and car sharing.

- Bicycle rental.

- Car-park use and park-and-ride systems.

- Admission to museums, musical events and sports events.

The field trial was conducted in close cooperation with the Verband Deutscher Verkehrsunternehmen (VDV), Cologne, Germany (www.vdv. de), which is now campaigning for introduction of a uniform smart card system throughout Germany. This is intended to be based on the VDV core application as a uniform system platform that would also form the basis for all future contractual projects. A corresponding research project involving renowned research institutes, industrial companies and the Deutsche Bank was recently concluded successfully [VDV2005]. The number of potential users of such a smart card system in Germany is enormous, with annual figures of 7 billion trips and 54 million passengers in public transport systems.

Approximately 5000 smart cards were issued initially in the Hanau field trial [RMV2005]. It was so successful that Volker Sparmann, CEO of RMV in Hofheim am Taunus, Germany, officially announced in a Hessischer Rundfunk (Hessen broadcasting corporation) programme broadcast in October 2005 that e-ticketing with smart cards would be offered to all users of the Rhein-Main local transport authority. NFC mobile telephones have since become a common means of payment in Hanau. The RMV is working to extend this innovation to its entire business area by around 2011 [RMV2006].

The contributors to the feasibility study generated prior to the Hanau pilot project included Deutsche Bahn AG (German Railways), DB Systems and other subsidiaries of Deutsche Bahn, as it can logically be expected that the smart card system tested there could be extended to railway transport.

10.5.3 New Active Smart Cards in Dresden

Another field trial of smart cards for local public transport was conducted in Dresden, Germany, in 2005. This involved a new generation of active smart cards equipped with an internal battery and two different RFID tags. The first tag operates at 7 MHz and has a maximum range of 3 m. It detects a signal field generated at the doors of Dresden buses and trams when they are open. When a passenger enters, the first tag activates the second tag in the smart card, and when the passenger leaves it deactivates the second tag. This extends battery life and avoids unnecessary tag use. The second tag is a UHF tag with an operating frequency of 868 MHz and a range of a several metres.

An antenna fitted inside each transport vehicle detects the activated UHF tags while the vehicle is underway and thus determines which smart cards are present in the vehicle. With this system, passengers do not have to swipe their smart cards past an antenna as in the Hanau system. Instead, they can keep them in their pockets. Fares are calculated automatically. This is called a 'be-in/be-out' (BiBo) method in contrast to the check-in/check-out (CiCo) method used in Hanau. Here again, mobile telephones were also included in the trial. Although RMV has already initiated system-wide use of smart cards, wide-scale implementation of the Dresden method will take several years.

10.6   Application Scenarios for NFC Mobile Telephones

NFC can be used for secure wireless communication over short distances (in the centimetre range) between electronic devices such as mobile telephones, PDAs, computers and automated payment devices. NFC thus competes with other noncontact connection technologies such as Bluetooth. The technical aspects of NFC communication and the NFC Forum (www.nfc-forum.org) formed by producers of technologies necessary for NFC are described in Chapter 7. Here we wish to present several application scenarios, some of which have been tested successfully in field trials or are already in operational use.

10.6.1   NFC Applications for Consumers

In April 2006, Royal Philips Electronics and Visa International published the results of a study on NFC technology for contactless payment, with the title 'How Would You Like to Pay for That? Cash, Card or Phone?'. According to the study, consumers rate the convenience, ease of use and 'coolness' of NFC mobile telephones very highly [Philips6604]. The study participants used NFC mobile telephones to make purchases in coffee shops, download films in DVD shops and buy concert tickets via smart posters (Figure 10.8), among other things.

In early 2006, Visa also announced that more than 20 000 POS in self-service restaurants, shops, cinemas and filling stations, predominantly in the USA, were already equipped with contactless payment terminals. Further expansion of NFC technology is thus very probable.

Figure 10.8   An NFC payment terminal at the Atlanta Stadium and a cinema poster with an NFC tag (sources: www.mobile-review.com, www.philips.at)

A smart poster has an RFID tag with a suitable marking. If an NFC telephone is held close to the tag, it can read the data stored in the tag. It can then call the ticket centre of a concert promoter or similar organization to reserve tickets immediately and pay for them. When the concertgoer arrives at the event location, he or she holds a mobile telephone next to the check-in terminal at the entrance and is immediately allowed to enter. For the concertgoer, standing in line to pick up a ticket is a thing of the past. It could hardly be faster or easier.

The publication describing the above-mentioned study also mentions other current pilot uses of NFC, such as by the Rhein-Main transport authority, where NFC mobile telephones have already become a normal means of payment (see Section 10.5). We can also mention the following pilot projects and scenarios here:

- **Philips Arena, Atlanta, USA.** Sports fans can use NFC here to buy products or download mobile contents – such as ring tones, screen savers or video clips – by holding their mobile telephone next to a smart poster.

- **City of Caen, France.** During a six-month trial project, 200 residents of the city used Samsung NFC mobile telephones to make payments in shops and car parks and download video clips, bus schedules and information about interesting sites.

- **Taiwan Proximity Mobile Service, Taiwan.** Prototype BenQ NFC mobile telephones were used to test secure payment transactions for public transport in Taiwan.

- **Ordering a taxi at a hotel.** An out-of-town hotel guest needs to order a taxi. In the future, the guest can simply touch his or her mobile telephone to the tag of a taxi poster in order to be connected to the taxi dispatch centre. The dispatcher knows the customer's location right away without asking and thus knows were to send the taxi.

- **Arriving in Paris by train.** A traveller is looking for the schedule of trains returning to the airport. The traveller touches the tag on the schedule and enters the desired departure time, and suitable connections are downloaded to his or her mobile telephone.

- **Paying parking fees and bus fares.** To pay for a parking ticket in a controlled parking zone, you can hold your mobile telephone next to a parking ticket dispenser. You can pay for a bus fare by swiping a mobile telephone past a check-in unit.

- **Access to ski areas.** Smart cards are already widely used in large ski areas for contactless access to the slopes. In the future, you can use your mobile telephone and avoid queuing at the lift station for a smart card. This also puts an end to arriving back home and discovering that you forgot to return the smart card and collect your deposit.

- **Calling a service hotline.** If you have a malfunction with a home appliance or trouble using the appliance and you want to sort this out by calling a service hotline, you usually do not know the telephone number. In the future, you could simply hold your mobile telephone next to the nameplate of the appliance and the call would be placed for you automatically.

10.6.2   Industrial Uses of NFC

NFC mobile telephones also offer new possibilities for industrial uses, since they are an order of magnitude less expensive than current RFID readers. They are thus a logical choice to replace RFID readers. This naturally assumes that employees who use the RFID readers already have mobile telephones. The 13.56 MHz NFC frequency is also widely used for RFID tags on individual objects (items), so the ROI in RFID methods can be accelerated considerably in trade and industrial applications.

For example, Nokia offers a service product for security and maintenance organizations that require full-coverage communication with their technicians or contractual partners. NFC mobile telephones can be used to manage and monitor the tasks of technicians and security personnel. For example, if the check points of a security round are fitted with RFID tags, the security guard only has to hold his or her mobile telephone next to the tags to document proper performance of the round in terms of locations and times. The same holds true for maintenance technicians if the equipment to be maintained is fitted with RFID tags. In this way, the necessary documentation of the work or the rounds can be generated automatically and concurrently with the activities.

10.7   Monitoring Components in Computer Centres

Computer centres usually house a bewildering number of components, such as racks, servers, routers, power supplies, and so on. The installed configuration of a computer centre changes rapidly. Studies show that one-third of the components are moved or modified each year to meet new requirements. If we assume a depreciation interval of five years, another fifth is replaced each year. On top of this, there are unplanned replacements due to repairs or failures. It is thus hardly surprising that it is usually difficult to keep track of the equipment inventory in a computer centre. The nearly omnipresent need for rapid implementation of changes is one of the most common causes of errors in inventory lists.

'Where is the server?' is thus a frequently heard question in computer centres, where stocktaking is not just an annual event but instead an almost daily activity due to constant configuration changes. When this question arises, it is a sign that regular computer centre activities associated with replacing, upgrading and repairing components are not documented as well as they should be. There is thus potential for improving the level of service or customer satisfaction in such computer centres.

This potential can be realized by implementing asset management based on a system management software package in combination with RFID tags and RFID readers.

10.7.1   Software Alone Is Not Enough

Modern system management software can be used to maintain lists of the exact locations of components, such as rack and slot numbers. However, this information is generally entered manually, and it is usually not up to date. In this respect, software-based lists are no better than handwritten lists.

Manual stocktaking is time-consuming, expensive and prone to error, and furthermore it is not performed often enough to ensure that the information is current. Using portable data entry units to acquire the data does not yield much improvement. Significant progress can be achieved by using barcodes, since this reduces the frequency of entry errors, but it does not yield a significant reduction in time expenditure – which is one of the main cost drivers. It also does not improve the currency of the data. An automated, near-real-time process for taking stock of the components in the computer centre is thus desirable.

Conventional automated approaches capture components in the computer centre that can be identified by their Internet (IP) addresses in the network. However, IP addresses do not contain any information about specific locations. In many cases, the location is ultimately determined by tracing the cabling, but this is usually difficult in practice. In any case, passive components such as power supplies, switches and monitors that are not connected directly to the Internet cannot be captured using the IP address method.

10.7.2   RFID to the Rescue

The necessary functionality can be achieved by using RFID. This involves fitting button-cell RFID transponders to the components, as illustrated in Figure 10.9. As many RFID antennas as necessary to precisely monitor the locations where components can be inserted in the rack are fitted to the rack door in the form of an antenna array. The transponders send their ID numbers to the antennas. Unlike the situation with barcodes, this occurs automatically and as often as desired, and it can also occur after a defined event such as closing the rack door. A sensor controller monitors the antennas in the rack and communicates the readout data to the central system management program, as shown in Figure 10.10. This

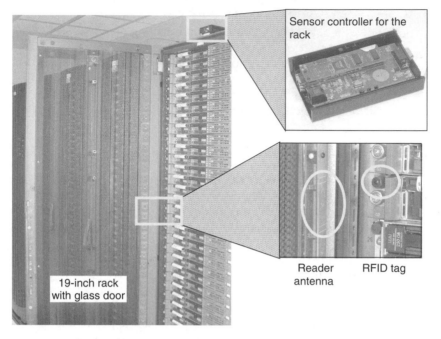

Figure 10.9 A 19-inch computer rack with an RFID antenna array (source: Hewlett-Packard)

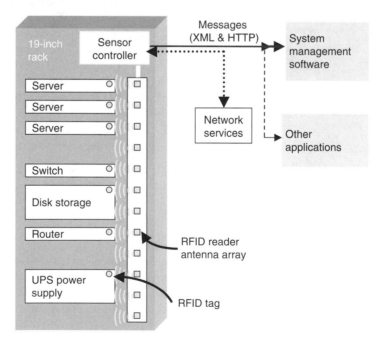

Figure 10.10 Communication between the software management system and the sensor controller (source: Hewlett-Packard)

system was demonstrated successfully as a research pilot for Hewlett-Packard server racks. In addition to collecting identification data, RFID readers can generate useful information by collecting data for parameters such as temperature and events such as opening and closing rack doors. Opening a door can signal an unauthorized access, so it should trigger an alarm (perhaps a silent alarm). Even without an alarm, removal of a component can be recognized by triggering an automatic scan after the door is closed and comparing the result with the existing inventory list. In any case, the assets database is updated, and it presents an accurate picture of the rack history and the components installed in the rack. The sensor controller and the management application communicate by exchanging messages in XML format via the network. An advantage of using rack-specific sensor controllers is that it makes identification of the components installed in a server rack independent of the following factors:

- network topology;

- server names;

- IP addresses;

- switched-on/switched-off status.

The RFID transponders used for this purpose are inexpensive and comply with ISO 15 693, and they are small enough to avoid interfering with air circulation in the rack. The short range of these transponders, which is often too small for other applications, is an advantage here because there will always be only one transponder within the reading range of each antenna, so its exact position can be recognized. A radiated power of 100 mW is fully sufficient to bridge the distance between the antenna and the transponder despite the difficult metallic environment and the electromagnetic interference inside the rack.

The RFID components form the basis for largely automated asset management. The ROI time is very short. Naturally, equipment outside the rack can also be identified independently of the sensor controller, for example by using a PDA fitted with an RFID antenna. This can also be used to acquire repair orders or document transfers to other departments. If a device is taken out of service, the system can automatically initiate cancellation of maintenance contracts and release of software licences for reuse.

Stepwise extension to other application areas, such as repair management or monitoring compliance with rules for minimum spacing or physical separation of components, can be achieved at low marginal costs.

Computer centres are usually operated as cost centres. One of the standard questions that must be answered with computer centres is whether the company is oversupplied or undersupplied with hardware and software. This question can only be answered if the equipment inventory of the computer system is known at all times. In addition, data regarding leasing, licensing and maintenance contracts, and warranty and service agreements must be captured and linked to user data. This is an essential prerequisite for accurate planning of the IT infrastructure, and with it the IT budget, and for examining service level agreements to identify opportunities for less costly arrangements.

Automatic identification of rack-mounted components is the first step toward comprehensive asset management, which can be followed by further steps. The asset management software infrastructure can gradually grow and develop into a core element of IT life cycle management, for example on the basis of Hewlett-Packard's Insight Manager product.

10.8   Aircraft Seat Quick-Change at Lufthansa

The operational profitability of aircraft depends on achieving the highest possible occupancy. Aircraft generate revenue when they are in the air and heading for the next airport with as many passengers as possible on board. On-ground times between arrival and departure and for maintenance work must be kept as short as possible. The occupancy level of the aircraft is an important key figure in the presentation of the annual results of an airline. Increasing this figure is thus an important objective.

For this reason, the seats in an aircraft are configured for long-distance flights to match the actual booking situation as closely as possible. No economy-class seats should remain empty if business class is overbooked, and vice versa. Consequently, at Frankfurt's Rhein-Main Airport the seating arrangements in the business and economy classes of Lufthansa's long-distance aircraft are adjusted during ground stops to match actual booking situations. This process is called 'passenger seat Quick-Change'. The following description is a summary from a thesis [Walther06].

The seat shop, which is located in one of the airport buildings, is responsible for this process. Its employees transport seats to the aircraft on trucks with hydraulic lift platforms ('highloaders'), reconfigure the seating and bring back the dismounted seats. The seat shop has three doors for loading and unloading the trucks. In the shop, the seats are inspected, stored and repaired as necessary. The number of seats held in stock varies and can be as high as 400 seat units (Figure 10.11).

The seats were previously identified using barcode labels, which had to be attached to the sides or bottoms of the seats where they could be seen by a barcode reader. As a result, labels attached in the passenger footroom area were often rendered unreadable by mechanical wear. By contrast, RFID tags can be attached in concealed positions where they cannot be damaged by unintentional actions but can still be read from above or below by service personnel without major effort.

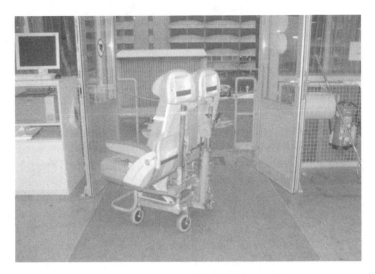

Figure 10.11 Quick-Change seating unit on a floor antenna in front of a loading door [Walther06]

Consequently, the seating units concerned were fitted with RFID tags for unique identification starting in 2005. This was also intended to speed up the Quick-Change process and eliminate errors in the generation of the legally prescribed documentation. After extensive testing, RFID tags with a transmission frequency of 13.56 MHz were selected. Although in principle the smallest possible tags should be used, tags with typical smart-card dimensions (76 × 45 mm) were chosen for superior readability over the required distance of more than 40 cm. It was also determined that the distance between the RFID tags and metal objects must be at least 25 mm, as otherwise the basic magnetic coupling process is not reliable. For this reason, a suitable holder made from Macrolon (a special plastic) was developed (Figure 10.12). The RFID tags and holders were specially tested and approved for this application by the development centre of Lufthansa Technik. Approval was not a problem from the electromagnetic perspective, since the tags are passive and are only

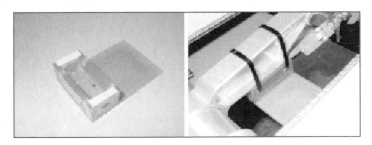

Figure 10.12 RFID tag with installation socket (left) and attached to a seat rail (right) [Walther06]

activated by the RFID antennas at the doors of the Quick-Change shop. Tests showed that they would not impair the onboard electronics systems of the aircraft.

The original intention was to store the product number (rail number; eight digits maximum) and serial number (12 digits maximum) of the seat in the RFID tag. However, it was found that this would cause unacceptable delays in the reading process at the shop doors. Consequently, it was decided to use the unique ID numbers stored in the chips during fabrication and employ a seat database on the server to associate them with the rail and serial numbers of the seats. Three shop doors for loading and unloading the trucks were fitted with RFID antennas to read the RFID tags. The antennas are recessed in the floor (marked by the dark area below the seat unit shown in Figure 10.11). The antennas are connected to the server.

The Quick-Change process consists of the following steps:

1. Before the airplane lands, the aircraft operations department informs the seat shop responsible for performing the Quick-Change of the new seating arrangement.

2. In the Quick-Change shop, the seats needed for the new configuration are loaded on a truck. The RFID antenna in the floor reads the ID numbers of the loaded seats. The software on the connected server enters the rail and serial numbers in the prepared installation orders. The truck then travels to the aircraft, where the installers reconfigure the seating arrangements in the aircraft.

3. The truck transports the dismounted seats back to the shop, and they are brought into the shop via a door. A floor antenna again recognizes the ID numbers and enters them in the dismounting order. This completes the process.

The server and the connected antennas are integrated into the IT network to make Quick-Change data available for processing by other systems. A supplementary benefit of this process is that an exact record of the total Quick-Change inventory is constantly maintained in central Lufthansa systems. It can provide answers to the following questions:

• How many seats are in the shop? Which of them need repair?

• How many seats are in the individual aircraft?

In principle, service personnel could also use a laptop computer fitted with a suitable RFID antenna to read the RFID tags in the aircraft. This possibility has not been used up to now. The completed Quick-Change process is still checked visually in the aircraft just as before.

This RFID-based system has been operating with 100% reliability since the pilot phase. The anticipated benefits have been achieved. It is certainly conceivable that the entire remaining inventory of seats (around 50 000) might sometime be fitted with RFID tags in order to improve the efficiency of asset management for them as well.

10.9 Trolley Asset Management at the Finnish Post Office

Trolleys are wheeled wire-mesh containers used to transport letters and parcels in post offices. The Finnish Post Office delivers 2.6 billion postal items per year, of which 26 million are parcels. It uses 200 000 trolleys for this purpose, and they cost around €300 each. The right number of trolleys must always be provided at the starting points of logistics routes, and they should be used as intensively as possible, as otherwise a larger stock of trolleys must be maintained than what would be necessary with economical use. Trolleys are an investment. If the number of trolleys can be reduced, this frees up capital for other uses, which has a positive impact on the bottom line. If trolleys are not available where they have to be filled with parcels, this quickly leads to additional employee effort and thus additional costs. Using software from BEA, the Finnish Post Office has created a system that supports these economic aspects.

Before this system was implemented, the post office was unable to monitor its trolley inventory reliably. It did not even know exactly how many trolleys it had. Shrinkage was also alarming; the cost of purchasing trolleys was more than €1 million per year. This was partly due to the fact that customers kept trolleys longer than necessary for the logistics tasks instead of returning them promptly to the post office. Costly shortages occurred regularly, especially during peak periods such as the Christmas season, and led to transport delays and customer irritation.

Naturally, the post office had experimented with barcodes, but barcode labels on such trolleys are exposed to mechanical wear and tear, so they are frequently unreadable or simply get lost. By contrast, RFID tags can be fitted in secure locations and recognized by RFID readers without visual contact. This is the solution to the problem.

Scalable, event-oriented middleware based on BEA's Weblogic Edge Server has been installed to process the data read by the RFID antennas from the RFID tags on the trolleys. The readers are placed at key points in the logistics route network of the post office, such as at depot entrances and exits, to improve the visibility of the supply process. Misrouted trolleys can also be recognized immediately and returned to the correct route.

The basis for this efficient asset management system for trolleys is a comprehensive, reliable database that was created during the pilot project and is now constantly expanded during processing of transport orders. The following data is acquired:

- trolley number, driver, delivery customer (recipient), and delivery date and time;

- trolley loading, unloading and transfer times.

This creates an accurate overview of trolley circulation, the number of transfers (events) per customer or route, daily and weekly trolley demand, and demand during peak periods.

It has been clearly shown that the RFID infrastructure allows costs (total cost of ownership), and in particular maintenance costs, to be reduced and trolley availability to be improved (the right number of trolleys at the right place and the right time). The visibility of logistics routes has also generally been improved, and locations where trolley losses typically occurred have been identified. The system also creates customer clarity because it always knows which trolleys are located with a particular customer, so correct invoicing is now possible in all cases.

RFID antennas are fitted at the entrances to the distribution centres (see Figure 10.13) to identify the trolleys during loading and unloading. In addition, portable RFID readers are used to identify trolleys no matter where they are. Special UHF tags operating at 868 MHz are attached to the trolleys. Up to 39 plastic boxes fitted with standard RFID tags can be transported in the trolleys. Both types of tags are recognized by the RFID readers with a 100% success rate. A rollout in the entire Finnish

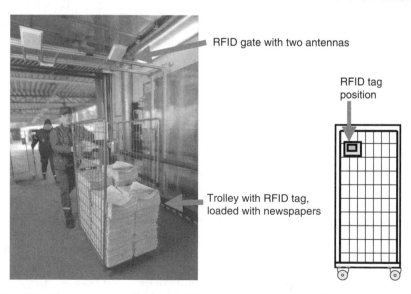

Figure 10.13 A wire-mesh trolley with RFID tags passing through an RFID antenna gate of the Finnish Post Office (source: BEA)

Post Office system has already started. Future plans envisage fitting every terminal of the post office with RFID readers and attaching tags to all delivery items, including bundled newspapers and individual letters. This step-by-step process will create increased visibility of postal routes for the post office and its customers.

10.10 Tracking Containers and Products at KPN in the Netherlands

The Dutch telecommunication enterprise KPN is involved in several experimental RFID projects. Its objective is to use the experience gained from these projects as a basis for developing managed IT service products that support enterprises in using RFID methods and exchanging real-time data with their trading partners. Two of these projects are described below:

- Tracking vegetable boxes fitted with RFID tags in the operational activities of Schuitema, a wholesale and retail organization that operates more than 450 C1000 supermarkets in the Netherlands.

- Bidirectional mobile telephone logistics between TNT distribution centres and KPN retail shops and business centres. Here the RFID tags are used as 'travelling databases'.

10.10.1 Tracking Vegetable Boxes at Schuitema

In December 2005, Schuitema launched a pilot project for using RFID tags with vegetable boxes. For this purpose, RFID antenna gates were set up at the entrances and exits of a distribution centre and the storage area of a supermarket (Figure 10.14). Heemskerk, a supplier of fresh vegetables, delivers the returnable vegetable boxes to the distribution centre. The boxes are identified by antennas when they arrive and leave the distribution centre and when they arrive at the C1000 supermarket.

KPN provided the IT infrastructure, and it operates the IT application system and the network. Its objective is to develop a managed service for RFID applications that can be provided to many enterprises in the supply chains of various industries. This is intended to achieve an economic benefit of scale that is not possible if each enterprise operates its own system. The KPN service allows Schuitema and the logistics service provider to concentrate on their core businesses and entrust data communication to KPN.

This is Schuitema's first experiment with electronic tracking of containers shipped from a supplier to a distribution centre and then to a

Figure 10.14 A pallet with vegetable boxes in an RFID antenna gate (source: Intellident)

supermarket. The system makes the data available to Schuitema as well as its suppliers. It collects EPC data in real time, processes the data and forwards the data. This is also the first EPC-oriented initiative of the Ahold Group trading organization. It primarily involves transporting fresh foods and vegetables. Schuitema also wishes to determine whether EPC-compliant Generation 2 technology is sufficiently robust to allow it to be extended to other supply chains.

The project is based on the following components from various companies:

- VI SixD, a managed service software program from VI Agents, which accepts and forwards RFID data along the supply chain.

- RFID readers from Feig Electronic.

- Vision software from Intellident, which acts as an edgeware component that processes the data written to or read from RFID tags and transmits the data to VI SixD.

- RFID tags from UPM Rafsec, with an operating frequency of 868 MHz (UHF) and chips from Philips (Ucode EPC).

Based on their experience, the suppliers of the RFID equipment plan to develop a general-purpose RFID antenna gate that can be preconfigured for use in other application scenarios and with other types of objects fitted with RFID tags, including objects with high concentrations of liquids and metals. This is intended to reduce on-site installation and operating costs.

10.10.2 KPN/TNT Tracking System for Mobile Telephones

KPN operates 100 retail shops and 17 business centres that sell mobile telephones. In 2006, KPN initiated a second RFID project called 'Telli-trace' in which it plays two roles: as a user and as a service provider. In this project, tags are attached to the boxes of individual telephones so they can be identified automatically when they leave the distribution centres operated by KPN and arrive at KPN businesses. The tags are removed at the POS. In addition, employees attach tags to unsold telephones to improve return logistics.

In this project, KPN and TNT are collaborating with RFID equipment supplier Symbol, which provides the readers and EPC Generation 2 tags (UHF) with 96-bit memory capacity, and Zebra Technologies, which provides the RFID tag printers. KPN hopes that the impact of RFID technology on its business processes will make this project a reference case from which all four companies can benefit.

Mobile telephones are small, valuable and coveted. If they are transported using normal logistics processes, there is considerable shrinkage. KPN presently operates screened 'warehouses inside warehouses' and uses separate transport processes for the telephones. In the future, when individual packages can be tracked using RFID, KPN intends to dispense with separate transportation and employ the processes used for other valuable products, such as cameras.

The pilot project calls for installation of six RFID readers and five RFID read/write stations in the supply chain, extending from the TNT distribution centre to two KPN Primafoon retail shops and two business centres. The RFID tags are attached to the boxes when the telephones arrive at the distribution centre and then read at the following locations:

- when they are stored in or retrieved from the screened security area;

- when they are picked for delivery and packed;

- when they pass by an RFID reader on a conveyor belt;

- when they leave the distribution centre;

- when they arrive at the stockroom of the retail shop;

- when they are taken from stock in the shop before tag removal at the POS.

The same tags are used for returning unsold telephones, for which purpose they are written with special KPN identification numbers. Codes identifying the reason for return, such as the wrong colour or a particular type of defect, are also entered. These tags are read when the returned items arrive at the TNT distribution centre. From there, nonfunctional telephones are sent for repair and functional telephones are put back in the supply chain or returned to the manufacturer. The aim is to store comprehensive information in the tags of the returned items so that all companies concerned receive information about the product without any need for a higher-level information system. This turns the tags into 'travelling databases'.

10.10.3 Accumulated Experience

KPN has accumulated a wealth of experience with all sorts of technical problems that must be solved before a satisfactory RFID read rate can be achieved. For instance, the 7-inch baskets for fresh vegetables cause considerable difficulties with reading because their tags are too close together if they are stacked on pallets. Magnetic coupling between the cards also occurs, which makes them unreadable.

With the boxes for mobile telephones, the tag attachment point and the position of the box in the environmentally friendly packaging present critical problems. If the distance between two RFID tags is less than 1 cm, they interact and are unreadable. If a tag is aligned exactly parallel to a CD or DVD in the box, it is also unreadable because the CD or DVD screens the electromagnetic radiation.

Problems of this sort cannot be overcome by improving the reader design, since they are based on physical laws. The only possible solutions are to modify the business process and packaging specifications or use application-specific tags and attach them to the boxes in a well-defined manner.

KPN recommends the first option if it yields sufficient benefits. Of course, it takes time because the processes (such as in the distribution centre) must be restructured and optimized. This is less difficult if the application involves pure asset management in a fixed area, such as in a hospital or office building.

The second option represents a task for the packaging industry. It will probably respond with specific tags for individual types of products and packaging and direct integration of the tags in the packaging, as is already the case with barcodes. KPN no longer believes in the possibility of a single, standard tag, and it instead foresees a broad variety of tags communicating with the readers.

Another lesson learned is that there is a considerable gap between the business rules followed by enterprises and the extent to which they can be modelled in support software. KPN is working closely with VI SixD software supplier VI Agents to upgrade the software to meet its requirements. Here again, there is no possibility of a general-purpose solution.

An interesting aspect here is the business strategy of VI Agents, which from early on has focused on the typical requirements of trade supply chains and the close relationship between data generated using RFID and benefits to enterprises involved in process chains that span enterprise boundaries [Green2005].

10.11   RFID Optimizes Garment Logistics at Gardeur

Gardeur AG, with headquarters in Mönchengladbach (Germany), was founded in 1920. The company achieved sales of €93 million in 2005 from producing and selling high-quality brand-name clothing. Approximately three million clothing items were produced, of which 90% were trousers and jeans. The export share rose above 50% for the first time ever in 2005. There are currently around 600 shop-in-shops and corners, of which 264 are outside Germany. The major export markets of gardeur are the German-speaking neighbouring countries, Scandinavia, the UK, Ireland and Russia. The first six shops in China are also developing well. The number of employees in the group increased by 264 to 1571. Most of the new hires were in Tunisia, where a laundry was added to the group's production facility.

Flexible, transparent logistics are a decisive factor for the success of the enterprise. For this reason, gardeur started using RFID technology to improve its inventory management as early as mid-2004. RFID is now an integral part of all merchandise flow processes. The benefit is proven, and the system is so effective that even Kaufhof AG, a subsidiary of the Metro Group and a major gardeur customer, wishes to participate in the RFID system. However, according to current knowledge this would require conversion from the circulating RFID tags currently used by gardeur to disposable tags complying with the EPC standard.

10.11.1   Complementary Use of Barcodes and RFID Tags

Every garment produced by gardeur is still labelled with a barcode. The barcode tag is attached to the hanger, while the RFID tag is attached directly to the garment (see Figure 10.15). These tags are removed from the garments at the destination and reused. For this reason, the pilot project only required a one-time investment for tags in the amount of €200 000.

A handheld reader is used to read the barcode number and RFID tag ID number simultaneously and link them in the merchandise management system. The barcode number (EAN 13 standard) is also stored in the RFID tag.

10.11.2   Nearly 100% RFID Read Rate

After this, an employee pushes the complete shipping unit (such as the rolling garment rack shown in Figure 10.16) through a RFID antenna gate.

Figure 10.15   RFID tags on hanging garments (source: gardeur)

Figure 10.16   Garment rack shipping unit being read while passing through an antenna gate (source: gardeur)

All of the RFID tags are identified in seconds with a reading accuracy of 99.78% based on project statistics from 18 months of operation.

10.11.3   Special Customer Wishes Are Welcome

The production activities of gardeur are highly structured and geographically distributed with small lot sizes. The company sees itself as a service provider, and it accommodates the special wishes of its customers by producing special finish versions. Accurate, synchronized acquisition of product data at the item level for each order at the interfaces in the production and logistics chains is thus important.

10.11.4   Continual Delivery Rate Optimization

The software for the RFID system was implemented in cooperation with RF-iT Solutions GmbH in Graz (Austria). It documents the goods and communicates goods transfers (such as by issuing a despatch advice) between the Augustfehn production units and the central warehouse and distribution centre in Mönchengladbach. Notification of shipments from Tunisia is also provided by electronic despatch advice messages. Each incoming shipment is identified automatically by a RFID reader and compared with the despatch advice. Missing or excess items are detected and resolved immediately. This method is used at all five production sites.

10.11.5 RFID Use Approaching Economic Viability

The infrastructure, consisting of the RFID reader hardware, interface software, middleware (edgeware) and consulting, required an investment of €310 000. Against this there are annual expenditures of €90 000 for depreciation and amortization, imputed interest, internal system support and maintenance fees. There is also the cost of purchasing tags for an annual production volume of three million garments per year, which amounts to €400 000, for total expenses of €490 000. The tags currently cost 15 eurocents each, and the procurement cost of the previous non-RFID labels is deducted from this.

The cost savings are estimated at €115 000 per year. These arise from process improvements in goods receiving and despatching, inventory management and internal transfers, as well as at the production sites, in the logistics centre and in the central warehouse. In addition, gardeur states savings and turnover gains resulting from improved data quality as follows:

- Improved delivery rate: €100 000 per year.

- Fewer discounts: €170 000 per year.

- Reduced shrinkage: €30 000 per year.

In total, gardeur currently realizes annual savings of more than €415 000 against total annual expenses of €490 000. The RFID system thus operates at a loss with disposable tags. However, gardeur expects that tag costs will decrease and the tags in the associated supply chains of its suppliers can also be used beneficially. In addition, gardeur is convinced that its early introduction of innovative RFID technology has enabled it to acquire a competitive efficiency advantage in the fashion industry.

10.11.6 The Metro Group Is Interested

Because of these excellent results, Kaufhof is currently negotiating with gardeur to have the RFID tags be left on the garments and aligned to the EPC standard. According to a report in trade magazine *Lebensmittelzeitung*, a project manager of the Future Store initiative at Kaufhof Warenhaus AG said that the aim was to have the RFID tags pay for themselves over the entire process chain [gardeur2006]. This is an interesting variant on the behaviour of the major retailer, which (like Wal-Mart in the USA) is better known for mandating requirements in order to persuade its suppliers to use RFID tags. In this case, gardeur has taken the initiative as an innovative supplier.

10.11.7   The Desired Quid Pro Quo: Sales Statistics

The commercial success of such a project hinges primarily on the price of the RFID tags. According to Production Director Ballweg, gardeur is prepared to use disposable tags conforming to the EPC standard if they are actually used in the retail industry. For gardeur, break-even could be achieved at a tag price of 12 eurocents.

Additional improvement potential could be realized if retail customers provided a quid pro quo. For instance, gardeur expects Kaufhof and other retail enterprises to create distinctly better visibility of the part of the supply chain extending inside their organizations as far as the POS. Retailers are still very reluctant to communicate their sales figures to their suppliers. However, producers could significantly improve their production and supply planning, and thus achieve further cost savings, if they had information on sales figures. In gardeur's view, it is thus clear that with inexpensive tags and access to retail sales reports, using EPC-compliant disposable RFID tags would immediately yield a convincing return on investment.

10.11.8   Technical Details

The RFID tags used by gardeur operate in the high-frequency (HF) band at the internationally available frequency of 13.56 MHz, which is preferred in many industries for tagging individual items. In the pilot project, these tags are removed when the garments are delivered and then reused. However, the aim is to leave the tags attached to the garments in order to avoid the extra process steps associated with reuse. Consequently, the number of tags that must be purchased each year is the same as the number of clothing items produced, which is currently more than three million.

The tags are currently written with a EAN 13 code, a gardeur-specific serial number and an ID number generated by the chip manufacturer. Changing to EPC would mean that the EAN 13 code and serial number would have to be changed to EPC format. This would also require making changes to gardeur's merchandise management system, although the company is willing to do this.

The proven read rate of the tags is 99.78%. The tags are used in five plants: three in Tunisia and two in Germany. The hardware and software are provided by Infineon, Sokymat, Schreiner Logidata and Logistic Ident Solutions, among others. Project implementation and integration were carried out by RF-iT Solutions (www.rf-it-solutions.com). Additional information is available at: www.gardeur.com.

10.12   Agent Technology in Intralogistics Systems[1]

The Fraunhofer Institute for Material Flow and Logistics in Dortmund, Germany, operates a conveyor system in order to develop and test new agent technologies in the field of intralogistics (processes in 'closed' systems inside enterprises). Some examples of the results of these research activities are described here. The associated technical infrastructure, which consists of agents that communicate via RFID and associated software platforms, is described in Chapter 6.

10.12.1   Controlling Continuous Conveyor Systems

Barcode labels are commonly used in many areas of logistics. Although barcode systems are presumably a more economical identification technology, currently published case studies [IDtech2512] demonstrate the higher economic and technological potential of RFID in logistics systems.

10.12.2   RFID: More Than a Barcode

Passive RFID tags and barcode systems have comparable use characteristics. In both cases, information about the physical material flow is synchronized at fixed locations with virtual inventory information in a logistics IT system. Scanners or readers are installed at these nerve centres in the material flow. Data is read where goods enter or leave the system by using stationary readers next to automated conveyor sections or portable readers.

If the data content remains the same, RFID tags can thus replace barcode labels. RFID tags have the advantage that the data contents of all tags in the antenna field of an RFID reader can be acquired concurrently and without physical or visual contact. The data content of a tag, which is an ID code assigned by the user, can be processed by the higher-level IT systems of an enterprise in the same way as a barcode ID.

The difference between barcode labels and RFID tags is that the latter are rewritable. Item information and process information in a tag can be modified, and the updated information will be available at every identification point along the logistics chain. This eliminates the need to access a WMS or ERP database when the tag is read. Currently available tags already have enough memory capacity to locally maintain a large amount of product-specific data and a complete process history.

10.12.3   Decentralization

With the memory capacity of such RFID tags, it is already possible to define process data for subsequent processing steps in advance and store

---

[1] Contributed by Dirk Liekenbrock, Fraunhofer Institute IML, Dortmund, Germany. www.iml.fraunhofer.de

Figure 10.17   Containers with RFID tags on a conveyor system (source: Fraunhofer IML)

the data in the chip of a container or pallet ('data on tag' method). Knowledge of the individual steps in the logistics chain is thus held by the chip instead of the higher-order IT system.

A natural application for this is controlling a container conveyor system (see Figure 10.17). Here an RFID tag attached to each container stores routing information about the route the container will take from were it enters the conveyor or production system to the previously calculated point where it leaves the system. With a complete network graph, alternative routes and changes to the operational process due to malfunctions or queues can also be taken into account. Read/write devices that can read the routing data is installed at all decision points (nodes) in the conveyor system. Each reader is connected to a controller that independently controls all conveyor and diverter drive and actuation mechanisms of the associated conveyor section.

The diverter controllers use a routing table to determine the direction setting for each identified container. The controllers also monitor the status of the individual sections still to be traversed by the container and can thus take malfunctions and queues into account for dynamic routing. Each time a container passes a decision point, a time stamp is written to the container chip after it is read. At the next decision point, the time stamp stored when the container was emptied, the time stamp written at the last node and the target time of arrival at a merger point, which is also stored in the chip, can be used to determine the handling sequence if two containers arrive at the same time.

Each controller can determine static routing entirely locally. All controllers in the system are connected by a network and can thus exchange system data with each other. Dynamic routing components require knowledge of the occupancy of subsequent sections of the route in other regions of the system. Here again the routing data stored in the container chips are helpful. The controller of the region where the chip is read uses this data to determine which other controller it must contact via the network to request occupancy data for that controller's region. These specifically requested data provide the basis for dynamic, load-dependent routing decisions.

The ability of the controller and the controlled section to handle real-time events also arises from the processes described above, since the communication overhead for control decisions is low because the controller only has to address the communication partners in the network that are needed for a particular container.

As each node makes control decisions on its own based on locally available information and can independently acquire other data as necessary, the result is a fully decentralized control architecture . The system layout and communication relationships, and with them the interoperability of the nodes as a whole, can be configured using XML files in a simple procedure when the system is put into service. It is fundamentally possible to extend this software architecture of the controllers to include a service-based architecture. In this case the route information in the RFID tags, which is stored in the form of a node graph, can be used to address Web services available in the network that relocate computation of control decisions from the decentralized controllers to a commonly available service.

A characteristic feature of such applications is the binding of individual items with the information necessary for the material flow. Each item can carry all of its descriptive data in an attached RFID tag. It is thus essentially possible not only to store and read information related to the current position of the item in the material flow, but also information related to subsequent process steps.

10.12.4   Agents for High-Bay and Small-Parts Warehouses

High-bay warehouses (HBWs) and automated small-parts warehouses (ASPWs) use a different type of conveyor system. These storage facilities are constructed as rows of shelves and aisles in which automated storage and retrieval units with load platforms for pallets or containers are used to store and retrieve items held in storage bays. The automated storage and retrieval unit travels horizontally along the aisle, while a load platform moving along vertical guides provides access to the various levels of each aisle. This arrangement is typical for warehouse facilities, and it harbours the risk that an entire aisle can be blocked if the associated automated storage and retrieval unit or one of its components fails. The items stored in the affected aisle are thus inaccessible for the duration of the failure.

Another possible manner to implement storage and retrieval of goods in storage bays is to use rail-guided warehouse vehicles (Figure 10.18) that travel along tracks at each level of an aisle and can be moved to a different level by a lift at the end of the aisle. The transfer capacity in the aisle can be scaled to match the warehouse throughput requirements by

Figure 10.18 Warehouse rail vehicle in an aisle with small-load carriers (source: Fraunhofer IML)

adjusting the number of vehicles assigned to the aisle. If a vehicle fails, the entire aisle is not blocked; in the worst case, only the level is blocked where the failed vehicle is located.

With this concept, aisle storage and retrieval of containers and transfer operations can be distributed over several vehicles and performed in parallel. The vehicles can be controlled using a wireless interface for transferring destination data from a control system to the vehicles and reporting back vehicle positions and current task status. The RFID-tagged containers in the warehouse can be identified using RFID readers fitted to the vehicles. The container ID in the tag can be used to check that the right container is being retrieved.

For execution of storage, retrieval and relocation tasks, the tasks must be assigned to the vehicles such that vehicle travel at a particular level and use of the vertical lift do no lead to collisions. Task assignment and monitoring is thus a key element of the associated task management system.

10.12.5   Association with an Agent System

Agent systems are suitable for performing task assignment and vehicle coordination functions. In this case, an agent instanced in the control system is associated with each vehicle. Tasks are published on a blackboard visible to all vehicle agents. Agents that do not currently have an assigned task use a bidding scheme to negotiate task assignments in order to achieve the most uniform possible utilization of the vehicles and time-optimized task execution. As an alternative to the central blackboard architecture of current systems, complete decentralization using active RFID components is also technically feasible. In this case, the agent software can be instanced directly in the RFID components of the vehicles.

11

Appendix

11.1 References

[Alb07] E-mail interview with Katherine Albrecht, 24 September 2007

[Anan2005] Raman, A., Narayanan, V. G.: Anreizsysteme: Teile und gewinne, *Harvard Business Manager*, September 2005, pp. 34–38. www.harvardbusinessmanager.de

[Arn95] Arnold, D.: *Materialflusslehre*, Vieweg, Braunschweig, 1995. www.vieweg.de

[Art05] Article 29 Group, Working document on data protection issues related to RFID technology, 10107/05/EN WP 105. http://ec.europa.eu/ justice_home/fsj/privacy/docs/wpdocs/2005/wp105_en.pdf

[Au06] Australian Law Reform Commission Issues Paper 31, Review of Privacy, No. 11.56, 2006. http://www.austlii.edu.au/au/other/alrc/ publications/issues/31/11.html#Heading210

[Auf06] Empfehlung der obersten Aufsichtsbehörden für den Daten-schutz im nicht-öffentlichen Bereich: Die Entwicklung und Anwendung von RFID-Technologie ist insbesondere im Handel und im Dienstleis-tungssektor datenschutzkonform zu gestalten!, 8–9 November 2006. http://www.bfdi.bund.de/cln_030/nn_531474/DE/ Oeffentlichkeitsarbeit/Entschliessungssammlung/DuesseldorferKreis/ November06RFID.html__nnn=true

[Ber2005] Berlecon Report: *RFID im Pharma und Gesundheitssektor*, Berlecon GmbH, Berlin, June 2005. www.Berlecon.de

[Bitkom2005] Hansen, W.-R. (ed.): *RFID White Paper*, BITKOM eV, Berlin, August 2005. www.bitkom.org (downloadable)

[Bleisch2003] Bleisch, G.: *Lexikon der Verpackungstechnik*, Hüthig, 2003

[Bode2004] Bode, Wolfgang: Lecture on packaging logistics presented at the Technical University of Osnabruck, 2004

[Boen2007] Boen, J.: Managed service initiative for auto ID/RFID applications in the airline industry, presentation at the VDI Airport Logistics
Conference, Nuremberg, Germany, 6 December 2006

[BSI2004] BSI study: *Risiken und Chancen des Einsatzes von RFID-
Systemen, Trends und Entwicklungen in Technologien, Anwendungen
und Sicherheit*, Bundesamt für Sicherheit in der Informationstechnik
(BSI), 2004. www.bsi.bund.de

[Bt07] German Bundestag Publication 16/4882; German Bundestag Publication 16/9248, p. 3. http://dip.bundestag.de/btd/16/048/1604882.pdf

[bvl2005] Straube, Pfohl, Günthner, Dangelmaier: Trends und Strategien
in der Logistik – Ein Blick auf die Agenda des Logistik-Managements
2010, Bremen 2005, *DVZ*, Deutscher Verkehrsverlag, Hamburg.
www.dvz.de

[CE81] Council of Europe, Convention for the Protection of Individuals
with Regard to Automatic Processing of Personal Data.
http://conventions.coe.int/Treaty/EN/Treaties/Html/108.htm

[Dan2003] Dangelmaier, W.: Grundlagen der Modellierung von Produk-
tionssystemen, lecture text, University of Paderborn, 2003

[ebus05] *The European e-Business Report 2005*, European Commission, Enterprise & Industry Directorate General, November 2005.
www.ebusiness-watch.org

[Eicar2006] Niedermeier, R. (ed.): *Leitfaden: RFID und Datenschutz*,
2006. www.eicar.org

[Enai05] Enaikoon Smart Box, product information of Enaikoon GmbH,
Berlin, July 2005. www.enaikoon.de

[EPC2005] http://www.epcglobalinc.org/public/ppsc_guide/

[EPC2503] *The EPCglobal Architecture Framework, Final Version*, 1 July
2005. www.epcglobalinc.com

[EU95/02] Data Protection Directive 95/46/EC, Official Journal L 281, 23
November 1995, p. 31; Electronic Data Protection Directive 2002/58/
EC, Official Journal L 201, 31 July 2002, p. 37

[EUCo07a] Commission of the European Communities, Commission Staff
Working Document, Results of the public online consultation on future
radio frequency identification technology policy, SEC (2007) 312,
p. 18. http://ec.europa.eu/information_society/policy/rfid/doc/rfidswp
_en.pdf

[EUCo07b] Communication from the Commission to the European
Parliament, the Council, the European Economic and Social Committee, and the Committee of the Regions: Radio frequency identification (RFID) in Europe: steps towards a policy framework, p. 5.
http://ec.europa.eu/information_society/policy/rfid/doc/rfid_en.pdf

[Fer01] Ferber, J.: *Multi-Agent Systems: An Introduction to Artificial Intelligence*, Addison-Wesley, 1999. www.addison-wesley.de

[Fink2002] Finkenzeller, K.: *RFID Handbook*, 2nd edn, Wiley, 2003.
www.wiley.com

[Flei2005] Fleisch, E., Mattern, F. (eds): *Das Internet der Dinge*, Springer, Berlin, 2005. www.springer.de

[Flei2006] Fleisch, E.: Crossing the point of no return, presentation at the Omnicard Conference, January 2006. www.omnicard.de

[Floer04] Floerkemeier, C., Schneider, R., Langheinrich, M.: Scanning with a purpose – supporting the fair information principles, *RFID Protocols*, p. 5. http://www.vs.inf.ethz.ch/publ/papers/floerkem2004-rfidprivacy.pdf

[Foe07] FoeBud, http://www.foebud.org/rfid/blockertags

[Fraport2005] Fraport spart 450 000 Euro pro Jahr, interview with Roland Krieg, CIO of Fraport AG, Frankfurt, *Computerwoche* 25, 24 June 2005, p. 24. www.computerwoche.de

[Gab2004] Krieger, K. (ed.): *Gablers Lexikon der Logistik*, 3rd edn, Gabler, Wiesbaden, 2004. www.gabler.de

[Gaja2002] Gaja, Lintner, Vogel: Dancing with the 800-Pound Gorilla, Boston Consulting Group, 2002. www.bcg.com

[Gam99] Gambardella, L. M., Taillard, E., Agazzi, G.: A multiple ant colony system for vehicle routing with time windows, Corne, D., Dorigo, M., Glover, F. (eds.), *New Ideas in Optimization*, McGraw-Hill, 1999, pp. 63–76. www.mcgraw-hill.com

[gardeur2006] Weber, B.: RFID im Tausch gegen Transparenz, *Textilzeitung*, Gardeur, 23 April 2006. www.gardeur.com

[GCI2005] *EPC: A Shared Vision for Transforming Business Processes*, GCI/IBM, G510-6200-00, 2005. www.gci-net.org

[Gil1994] Gillert, Külpmann, Mühlencoert: *Studie zur Quellensicherung*, Jansen, R. (ed.), Logistics Department, University of Dortmund, 1994. www.flog.mb.uni-dortmund.de

[Gil2001] Gillert, F.: Entwicklung einer Methodik zur labortechnischen Abnahme quellengesicherter Produkten und Produktverpackungen, Jansen, R. (ed.), *Transport und Verpackungslogistik*, vol. 52, 2001. www.flog.mb.uni-dortmund.de

[Green2005] Green, N. *et al.*: Turning signals into profits in the RFID-enabled supply chain, Doukidis, G. *et al.* (eds), *Consumer Driven Electronic Transformation*, Springer, New York, 2005. www.springeronline.com

[GS12005a] Tracking & Tracing, GS1-Standards sorgen für Transparenz, brochure, GS1 Germany GmbH, Cologne, 2005. www.gs1-germany.de

[Ham2003] Hambuch, P.: Collaborative planning, forecasting and replenishment, ms, Procter & Gamble, 2003

[Hansen06a] Hansen, W.-R.: RFID – Nutzen durch Erweiterung der IT-Infrastruktur, *Jahrbuch eBusiness*, Wegweiser Verlag, Berlin, 2006. www.wegweiser.de

[Hansen06b] Hansen, W.-R.: Wir sind noch nicht so weit, *RFID im Blick*, Amelinghausen, March 2006. www.rfid-im-blick.de

[Hardgrave05] Hardgrave, B. C. *et al.*: *Does RFID Reduce Out of Stock? A Preliminary Analysis*, University of Arkansas, 2005. http://itrc.uark.edu

[Harmon2006] Harmon, C.: The necessity for a uniform organisation of User Memory in RFID, *International Journal of Radio Frequency Identification Technology and Applications*, 1, no. 1, 2006. www.inderscience.com

[Hei07] heise news, 9 November 2007. http://www.heise.de/newsticker/result.xhtml?url=/newsticker/meldung/95797&words=RFID&T=rfid

[Hein2003] Heinrich, C.: *Adapt or Die: Transforming Your Supply Chain into an Adaptive Business Network*, John Wiley & Sons, Ltd, 2003. www.wiley.com

[Hein2005] Heinrich, C.: *RFID and Beyond: Growing Your Business Through Real World Awareness*, John Wiley & Sons, Ltd, 2005. www.wiley.com

[Helbing2003] Helbing, D.: Modelling supply networks and business cycles as unstable transport phenomena, *New Journal of Physics*, 5, 2003, p. 90. www.njp.org

[Hors2005] Horst, F.: Zehn Jahre ECR, *retail technology* 04/2005, pp. 10–13. www.ehi.org

[HR3580] House of Representatives, 3580 Food and Drug Administration Amendments Act of 2007, Title IX – Enhanced Authorities Regarding Postmarket Safety of Drugs – Subtitle B – Other Provisions to Ensure Drug Safety and Surveillance, Section 505d, 'Pharmaceutical Security'. http://thomas.loc.gov/cgi-bin/bdquery/z?d110:h.r.03580:

[IDtech2512] Summary of case studies in December 2005. www.idtechex.com/knowledgebase

[Jap04] Guidelines for privacy protection with regard to RFID tags, 8 July 2004. http://www.meti.go.jp/english/information/data/IT-policy/pdf/guidelines_for_privacy_protection_with_regard_to_rfid_tags.pdf

[Kel2004] Keller, S.: Die Reduzierung des Bullwhip-Effektes – eine quantitative Analyse aus betriebswirtschaftlicher Perspektive, dissertation, University of Duisburg-Essen, 2004

[Kleist2005] Kleist, R. A. *et al.*: *RFID Labeling – Smart Labeling Concepts & Applications for the Consumer Packaged Goods Supply Chain*, Printronix Inc., Irvine, CA, 2005. www.printronix.com

[Ko07] RFID privacy protection guideline, 2007. http://onepark.khu.ac.kr/

[Kok04] Kokozinsk, R., Holzapfel, M., Coers, A., Müller, H.: Drahtlose Sensornetzwerke – Technologie und Architektur, *Sensorreport* 6, 2004, St Gallen. www.sensorreport.de

[Kon06] Entschließung der 72, Konferenz der Datenschutzbeauftragten des Bundes und der Länder, Naumburg, 26–27 October 2006. http://www.datenschutz-berlin.de/doc/de/konf/72/RFID.pdf

[Laendm2007] *Programmierte Verfolgung mit Laendmarks*, project report, 2007. www.nextgenerationmedia.de

[Lan05] Langheinrich, M., in Fleisch, E., Mattern, F. (eds): *Das Internet der Dinge*, 2005, pp. 341 ff

[Lan08] Langheinrich, M., RFID und die Zukunft der Privatsphäre, in Rossnagel, A., Sommerlatte, T., Winand, U. (Eds.), *Allgegenwärtige Datenverarbeitung – Wie möchten wir in Zukunft leben?*, pp. 4, 28, 12ff. http://www.vs.inf.ethz.ch/publ/papers/langhein-rfid-de-2006.pdf

[Man2005] Mannel, A., Koch, R.: Beispielhafte Bewertung einer idealtypischen Prozesskette in der Textillogistik, Seifert, W., Decker, J. (eds), BVL, *RFID in der Logistik – Erfolgsfaktoren für die Praxis*, publication in the *Wirtschaft & Logistik* series, Deutscher Verkehrs-Verlag, Hamburg, 2005, pp. 143–168. www.dvz.de

[Man2006] Mannel, A.: Entwicklung eines Kostenbewertungsmodells für den Einsatz der RFID-Technologie in der Bekleidungsindustrie auf Basis betrieblicher Kennzahlen, AIF research project final report, Dortmund, 2006. www.aif.de

[Metro2005] Wolfram, G., Rindle, K.: *RFID – A Driving Force for Innovation*, Metro Group/IBM, Düsseldorf/Herrenberg, 2005. www.ibm.com

[Metro2603] Metro Group RFID newsletters, March 2006 and June 2006. www.future-store.org

[Moore96] Moore, G. A.: *Das Tornado Phänomen*, Gabler, Wiesbaden, 1996. www.gabler.de

[Mül05] Müller, H., Holzapfel, M., Coers, A., Kokozinsk, R.: Drahtlose Sensornetzwerke – Systeme und Anwendungen, *Sensorreport*, January 2005. www.sensorreport.de

[NY07] Section 219 (3) General Business Law as in New York 2007 Assembly Bill 222 (RFID Right to Know Act). http://assembly.state.ny.us/leg/?bn=A00222&sh=t

[OCon2006] O'Connor, M.: Intelleflex will supply a chip for the passive tags Boeing wants placed on parts for its Dreamliner jets, *RFID Journal*, 4 April 2006. www.RFIDjournal.com

[OECD80] OECD: Recommendations of the Council Concerning Guidelines Governing the Protection of Privacy and Transborder Flows of Personal Data, OECD-Document C (80) 58 (Final). http://www.oecd.org/document/18/0,3343,en_2649_201185_1815186_1_1_1_1,00.html

[Philips2604] *How Would You Like to Pay for That? Cash, Card, or Phone?*, study on near-field communication (NFC), Philips/Visa, 2006. www.nxp.com

[Pichler02] Pichler, J., Plösch, R., Weinreich, R.: MASIF und FIPA: Standards für Agenten Übersicht und Anwendung, *Informatik Spektrum*, 25, no. 2, April 2002. www.springer.com

[Progress2503] Trigg, J. B.: *Progress for RFID – An Architectural Overview*, Progress Software Corp., Bedford, MA, March 2005. www.progress.com

[Progress2508] *Progress: Three Components of SOA*, white paper, Progress Software Corp., Bedford, MA, August 2005. www.progress.com

[Rindle2607] Rindle, K.: Secure Trade Lane, presentation to Bitkom RFID Project Group, 5 July 2006

[RMV2003] *Electronic Ticketing in der Region Frankfurt-Rhein-Main*, feasibility study, Rhein-Main-Verkehrsverbund GmbH and Deutsche Bahn AG, 2003. www.rmv.de

[RMV2005] Ordon, C.: E-Ticketing im Rhein-Main-Verkehrsverbund, text of speech, Hanau, 2005. www.get-in.de

[RMV2006] Handy statt Fahrschein, *Frankfurter Allgemeine Zeitung*, 20 April 2006. www.faz.net

[SA06] Independent Communications Authority of South Africa, no. 1141, 08/25/2006, Notice in Terms of Section 46 of ICASA Amendment Act, no. 3 of 2006, inviting representations with regard to spectrum reallocation to cater for radio frequency identification technology. http://www.info.gov.za/gazette/notices/2006/29161.pdf

[Schm05] Schmid, V.: Mastering the legal challenges, in Heinrich, C. (ed.): *RFID and Beyond, Growing Your Business Through Real World Awareness*, 2005, p. 194

[Schm08] Schmid, V.: Radio Frequency Identification Law Beyond 2007, Internet of Things, *Springer Lecture Notes on Computer Science*, 2008

[Scho1999] Scholz-Reiter, B., Jakobza, J.: Supply chain management, *HMD* 207, 1999, pp. 7–15, ISSN 1436-3011. http://hmd.dpunkt.de/207/01.html

[Speck2004] Specker, A.: *Modellierung von Informationssystemen*, Vdf Hochschulverlag, Zürich, 2004. www.vdf.ethz.ch

[Staake2005] Staake, T., Thiesse, F., Fleisch, E.: *Extending the EPC Network – The Potential of RFID in Anti-counterfeiting*, Auto-ID Lab, St Gallen, 2005. http://vsgr.inf.ethz.ch/autoidlabs.ch

[STOA07] European Parliament, Scientific Technology Options Assessment STOA, RFID and identity management in everyday life, June 2007, pp. 15, 52. http://www.europarl.europa.eu/stoa/publications/studies/stoa182_en.pdf

[Strass2005] Strassner, M.: *Innovationspotenzial von RFID für das Supply Chain Management*, dissertation, University of St Gallen, 2005. www.unisg.ch

[Strau2005] Straube, Dangelmaier, Günther, Pfohl: *Trends und Strategien in der Logistik – Ein Blick auf die Agenda des Logistikmanagements 2010*, Bundesvereinigung Logistik, Bremen, October 2005. www.bvl.de

[Swee2005] Sweeney, P. J.: *RFID for Dummies*, John Wiley & Sons, Ltd, April 2005. www.wiley.com

[tenHo04] ten Hompel, M., Lange, V. (eds): *RFID – Logistiktrends für Industrie und Handel*, Verlag Praxiswissen GmbH, Dortmund, 2004. www.iml.fraunhofer.de

[tenHo05] ten Hompel, M., Liekenbrock, D.: Autonome Objekte und selbst organisierende Systeme – Anwendung neuer Steuerungsmethoden in der Intralogistik, *Industrie Management* 23, 2005. www.industrie-management.de

[Terat2005] TeraTron LPS im neuen Gotthard-Basistunnel, product information, Teratron GmbH, August 2005. www.teratron.de

[Tria04] Jahn, A.: GSM data support for container ships, presentation slides, Oberpfaffenhofen, 2004. www.triagnosys.com

[TStar07] Toronto Star, Device made in Ontario could save stuck miners. www.thestar.com

[UN90] UN guidelines concerning computerized personal data files, Res. 95/45. http://www.unhchr.ch/html/menu3/b/71.htm

[Unisys2004] *Your New World, A Visual Guide to Global Commerce*, Unisys, October 2004. www.unisys.com

[USFIP73] US Fair Information Principles, US Department of Health Education and Welfare, 1973. www.hhs.gov

[VDA2006] *RFID im Behältermanagement der Supply Chain*, Recommendation 5501 of the German Association of the Automotive Industry (VDA), Frankfurt, November 2006. www.VDA.de

[VDI] VDI Guideline 4470 (1999) and 4471 (2002). www.vdi.de

[VDV2005] VDV-Forschungsprojekt schafft elektronischen Ticket-Standard, VDV press release, Cologne, 27 July 2005. www.vdv.de

[Vog2004] Vogell, K., Kranke, A.: CRP, VMI und CMI, ECR series no. 9, *Logistik inside*, Munich, March 2004. www.logistik-inside.de

[Walther06] Walther, M.: *Einführung der RFID-Technologie bei der Pax Seat Versionsanpassung*, dissertation, Lufthansa Technik, Frankfurt, 2006

[Warb2004] Warburton, R. D. H.: An analytical investigation of the bullwhip effect, *Production and Operations Magazine*, 13, no. 2, summer 2004, pp. 150–160. www.poms.org

[Wass2007] Wasserman, E.: Airbus' grand plans for RFID, *RFID Journal*, October 2007. www.RFIDjournal.com

[Weis1991] Weiser, M.: The computer for the 21st century, *Scientific American* 265, no. 3, 1991, pp. 66–75. www.sciam.com

[Werle2005] Werle, K.: Aldi trifft Gucci, *Manager Magazin*, Hamburg, January 2005. www.managermagazin.de

[Wool02] Wooldridge, M.: *Introduction to Multi-Agent Systems*, John Wiley & Sons, Ltd. 2002. www.wiley.com

[WPC03] World's privacy regulators call for privacy friendly RFID tags, press release by the Privacy Commissioner of Australia, 2003. http://www.privacy.gov.au/news/media/03_17.html

11.2   Information Sources on the Internet

11.2.1   Universities, Associations, Committees and Manufacturer-Independent Institutions

www.aim-d.de	AIM section for Germany, Austria and Switzerland
www.aimglobal.org	Association for Automatic Identification and Mobility, Inc. (AIM), Warrendale, USA and Brussels, Belgium
www.airlines.org	Air Transport Association (ATA)
www.autoidlabs.org	worldwide network of Auto-ID Centres
www.bitkom.org	largest German IT industry association
www.bme.de	Bundesverband Materialwirtschaft, Einkauf und Logistik (BME)/Association of Materials Management, Purchasing and Logistics (AMMPL)
www.bsi.bund.de	Bundesamt für Sicherheit in der Informationstechnik/German Federal Office for Information Security (BSI), Bonn
www.bvl.de	Bundesvereinigung Logistik eV (BVL)/German Logistics Association, Bremen
www.chep.com	worldwide pallet and packaging logistics
www.ean-int.org	EAN International, Brussels, Belgium
www.ecrnet.org	ECR initiative under the sponsorship of GS1
www.eecc.info	European EPC Competence Center; GS1 testing and training centre in Cologne, supported by the Metro Group
www.eicar.org	EICAR eV, European expert group for IT security
www.epcglobalinc.org	EPCglobal Inc., a subsidiary of GS1
www.epic.org	Electronic Privacy Information Center (EPIC), Washington, DC
www.ero.dk	European Radiocommunication Office, Copenhagen (regulatory body)

www.faa.gov	Federal Aviation Administration, USA
www.fda.gov	Food and Drug Administration, USA
www.flog.mb.uni-dortmund.de	Logistics Department (FLog) of the University of Dortmund, and the associated LogIDLab logistics ID laboratory
www.foebud.org	Verein zur Förderung des öffentlichen bewegten und unbewegten Datenverkehrs eV (FOEBUD), Bielefeld, Germany
www.future-store.org	Metro Future Store Initiative
www.gci-net.org	Global Commerce Initiative, global industry initiative to promote cooperation in the trade sector
www.gs1.org	GS1 International, the international GS1 umbrella organization
www.gs1-germany.de	GS1 Germany GmbH, Cologne, formerly Centrale für Coorganisation (CCG)
www.ieee.org	Institute of Electrical and Electronics Engineers
www.iml.fhg.de	Fraunhofer Institute for Material Flow and Logistics (IML), Dortmund, and the associated openID centre
www.info-rfid.de	Informationsforum RFID, Berlin
www.internetderdinge.de	An initiative of the Fraunhofer Institute IML, Dortmund
www.ita-int.org	Informationstechnologie für die Automobilindustrie (ITA), an industrial initiative of the Verband Deutscher Automobilindustrie/German Association of the Automotive Industry (VDA)
www.item.ch	Institute of Technology Management, University of St Gallen
www.licon-logistics.de	logistics association sponsored by Kühne + Nagel, Siemens and others
www.logistics.about.com	logistics information
www.logistik-lexikon.de	logistics lexicon (German)

www.m-lab.ch	M-Lab, associated with the University of St Gallen and ETH Zürich
www.myLogistics.net	logistics portal of AXIT AG
www.nfcforum.org	international industrial association for promotion of NFC technology
www.nocards.org	Consumers Against Supermarket Privacy Invasion and Numbering (CASPIAN); US consumer rights organization
www.odette.org	Odette International Ltd, London, UK
www.openid-center.de	an initiative of the Fraunhofer Institute IML for construction of a networked value chain based on RFID technologies
www.rfid-handbook.de	Internet site of the *RFID Handbook* by Klaus Finkenzeller
www.uc-council.org	Uniform Code Council, USA
www.vics.org	Voluntary Interindustry Commerce Standards Association (VICS), Lawrenceville, NJ; sponsor of the CPFR initiative
www.w3c.org	World Wide Web Consortium
www.wcoomd.org	World Customs Organization (WCO), Brussels
www.wi-fi.org	WiFi Alliance, Austin, TX
www.wimaxforum.org	WiMax Forum, Mountain View, CA

11.2.2 Media

www.dvz.de	*Deutsche Verkehrszeitung/Deutsche Logistik-Zeitung*
www.eurocargo.de	trade magazine
www.eweek.com	trade magazine focusing on RFID
www.funkschau.de	trade magazine
www.gpsworld.com	US magazine focusing on GPS
www.ident.de	trade magazine published by the AIM section for Germany, Austria and Switzerland
www.ipm-scm.com	supply chain management trade magazine

www.logistik-heute.de	magazine published by BVL
www.logistik-inside.de	trade magazine with detailed RFID documentation
www.logistikmarkt.ch	Swiss logistics portal
www.rfidforum.de	trade magazine/smart card forum
www.rfid-forum.de	*ident* is the official publication of AIM in Germany
www.rfid-im-blick.de	German RFID trade magazine
www.rfidjournal.com	US RFID trade magazine, also organized conferences
www.scs-mag.com	supply chain trade magazine
www.telematicsjournal.com	*Telematics Journal*
www.themanufacturer.com	US trade magazine

11.3   Glossary

AFI	Application Family Identifier (data field of the ISO 18000-6 standard)
Air interface	Electromagnetic field used for data transmission between RFID reader antennas and tag antennas
AIP	Air interface protocol, e.g. Generation 2
ALE	Application Level Event; interface between RFID middleware and applications, specified by EPCglobal
ASCII	American Standard Code for Information Interchange (seven-bit code, originally developed by the American National Standards Institute)
ASN	Advanced Shipping Notification
Auto-ID Labs	Research institutions at MIT (Boston, USA), the University of St Gallen (Switzerland), Cambridge University (UK) and other locations
Auto-ID	Automatic noncontact identification of objects; general term for methods such as barcodes and the RFID research area of Auto-ID Labs
BDI	Beliefs, desires and intentions (term used in agent architectures)
BPM	Business process management
CAGE	Commercial and government entity (term used by US government bodies and the DoD)

CASPIAN	Consumers Against Shopping Privacy Invasion and Numbering
Chip	Electronic data storage medium and processor in an RFID tag
CORBA	Common Object Request Broker Architecture
CPFR	Collaborative Planning, Forecasting and Replenishment, ECR concept for cooperative demand planning in the trade sector, devised by GS1
CPG	Consumer packaged goods; refers to the consumer goods industry (*see* FMCG)
DESADV	EANCOM despatch advice message
DI	EAN 128 data identifier
DIY	Do it yourself; e.g. DIY software as opposed to standard software
DMS	Document management system
DNS	Domain name server; Internet directory for converting WWW addresses into IP addresses
DoD	US Department of Defense
DODAAC	DoD Activity Address Code; packing list number used in military sector
DSL	Digital Subscriber Line
DTV	Driverless transport vehicle in a production system
DUNS	Data Universal Numbering System; an internationally recognized 9-digit numerical code for unique identification of enterprises, devised in1962 by Dun & Bradstreet
EAI	Enterprise application integration; function of modern middleware in an IT architecture
EAN	European item number in 13-character barcodes; counterpart to UPC
EAN 128	GS1 EAN format combining data formats and data identifiers such as item number, batch number, etc.; can represent the entire ASCII character set.
EAN International	Standardization organization; merged into GS1 International
EANCOM	Formatting standard for EDI data, specified by GS1
EAS	Electronic article surveillance; RFID method used in the retail sector for theft prevention

EBCDIC	Extended Binary Coded Decimal Interchange Code; an 8-bit character encoding scheme originally used with IBM mainframe operating systems
ECR	Efficient Consumer Response; an enterprise initiative in the trade sector in cooperation with GS1
Edgeware	Specific middleware for controlling and monitoring RFID readers. The EPCglobal specification for this is called Savant
EDI	Electronic data interchange; electronic messages in logistics chains, e.g. despatch advice
EGNOS	European Geostationary Navigation Overlay Service; European satellite system for enhancing GPS signals
EHIBCC	European Health Industry Business Communication Council; European body that issues HIBC codes
EICAR	European Institute for Computer Anti-Virus Research; also maintains an RFID task force
EPC	Electronic Product Code, consisting of a manufacturer number, product number and serial number issued by GS1
EPCglobal	Nonprofit organization that promotes RFID research and commercial use of RFID; belongs to GS1 International
EPCIS	Middleware component for access to the EPCglobal network
EPOSS	European Technology Platform On Smart Systems Integration, Brussels; special-interest group of European industrial enterprises such as Siemens, EADS, Infineon and Metro
ERP	Enterprise resource planning; abbreviated designation for software systems that support the commercial tasks of enterprises; implemented as custom or standard software or a mix of the two
ETSI	European Telecommunications Standards Institute
EUI- 64	Electronic code for use on RFID tags; specified by IEEE
Event	An occurrence in an auto-ID system, such as reading an RFID tag at a warehouse entrance. This causes the IT system to receive data, which must be forwarded.
FAA	Federal Aviation Administration (USA)
FDA	Food and Drug Administration (USA)

FMCG	Fast-moving consumer goods; designation for the retail sector
Galileo	European civil satellite-based positioning system, currently under construction; future competitor for GPS
GCI	Global Commerce Initiative; promotes commercial use of electronic methods under the direction of IBM and other suppliers
GDA	General association of the aluminium industry
GDS	Global Data Synchronization; service offered by GS1
GDSN	Global Data Synchronization Network; network for providing uniform item master data under the direction of GS1; 450 member enterprises as of 2005
Generation 2	New tag standard for the EPCglobal air interface, submitted to ISO for standardization; abbreviated as Gen-2
GIAI	Global Individual Asset Identifier; asset number used for objects such as hospital beds, computers and delivery vehicles
GLN	Global Location Number; EPC for companies, sites, warehouses, etc.
GLONASS	Global Navigation Satellite System; operated by Russia
GNSS	European Global Navigation Satellite System for civil use; synonym for Galileo
GPC	Global Product Classification; GS1 service
GPRS	General Packet Radio Switching; virtual private network technology used in GSM networks
GPS	Global Positioning System (see also Galileo)
GRAI	Global Returnable Asset Identifier consisting of a manufacturer number and a serial number; used for trailers, railcars, beer kegs, gas cylinders, etc.
GS1 Global Registry	Central network authority in GDSN
GS1 International	Global standardization organization in the trade sector (formerly EAN and UCC) with member organizations in more than 100 countries
GSM	Global System for Mobile Telecommunication
GSMP	Global Standards Management Process (GS1 initiatives)
GTIN	Global Trade Item Number; EAN number with 8, 12, 12 or 14 digits containing a manufacturer number and object class (see SGTIN)

HIBC	Health Industry Barcode; European standard devised by EHIBCC
HTML	Hypertext Markup Language; formal language used to design websites
HTTP	Hypertext Transfer Protocol; used to publish HTML contents addressed by URLs
IEEE	Institute of Electrical and Electronics Engineers; adopted a standard for electronic codes called EUI-64
IFA	Informationsstelle für Arzneimittelspezialitäten GmbH; issues PZN codes in Germany
Inlay	Unfinished version of an RFID tag, consisting of a chip and an antenna, for used with labels, smart cards, etc.
Interrogator	Synonym for reader (e.g. RFID reader)
INVOIC	Electronic invoice sent via EDI (EANCOM)
ISO	International Organization for Standardization
Item	An individual product; the smallest unit for application of an RFID tag, as opposed to a case or pallet
ITIL	IT Infrastructure Library Guidelines developed on commission of the British government; worldwide de facto standard for service management
JIS	Just in sequence; an organizational principle in the automotive industry
JIT	Just in time; an organizational principle in the automotive industry
KPI	Key performance indicators; used to measure performance
LICON	Logistic Ident Consortium; a German association that promotes the use of RFID tags in logistics systems, operating under the direction of Kühne+Nagel and Siemens Business Services
M2M	Machine-to-machine communication, facilitated by auto-ID methods such as RFID
MAS	Multiagent system, such as for controlling conveyor belts
MES	Manufacturing execution system
Middleware	Software level in the middle (integration) layer of a software architecture

NFC	Near-field communication; RF data transmission method with an operating frequency of 13.56 MHz and a maximum reading distance of 20 cm; ISO standard; originally developed by Philips and Sony
NVE	German designation for EAN 128 shipping unit numbers used to identify pallets, etc.; equivalent to SSCC
Object website	'Website' of an object (item, case, pallet, etc.) on the internet of things; the associated IP address is provided by ONS
ONS	Object Name Service; central directory of EPC numbers for the EPCglobal network and associated IP addresses of object websites
ORDERS	Order placed using EDI; EANCOM standard
PLM	Product life cycle management
PML	Physical Markup Language; XML variant used for communication in EPC networks
POS	Point of sale; electronic cashier system in department stores
Proximity	General term for near-field technologies such as NFC, RFID and Bluetooth; *see also* Vicinity
PZN	Central pharmaceutical number used in Germany; issued by IFA
RCS	Radar cross-section, an imaginary surface area used in radar technology to characterize the reflectivity of objects. It is used in the RFID context to specify the reflectivity of tags.
RECADV	Receiving advice message (EANCOM)
RFID	Radio frequency identification
RFID tag	A wireless label consisting of a chip with a processor and data storage combined with an antenna for data transmission using radio signals (RF)
ROI	Return on investment
RPC	Remote procedure call
RSS	Reduced Space Technology for barcodes (EAN/UCC specification)
RTI	Returnable transport items (containers, crates, etc.)
RTLS	Real-time locating system
RWA	Real-world awareness, a term promoted by SAP, based on auto-ID products and systems

Savant	Edgeware component specified by EPCglobal
SCM	Supply chain management; a component of or complement to ERP
SGTIN	Serialized Global Trade Identification Number; a GTIN with the addition of a serial number for identification of individual objects
SHF	Super high frequency (microwave)
SKU	Stock-keeping unit; a package or case
SLA	Service level agreement; a contract governing services and associated measurement criteria in the form of key figures that describe the promised level of performance
Slap and ship	Colloquial designation for initial use of RFID: generating an RFID label (including barcode printing), 'slapping' it onto a pallet and shipping
Smart chips	Synonym for RFID tags
SNA	System Network Architecture (IBM)
SOA	Service-oriented architecture, in particular a property of flexible middleware
SOAP	Simple Object Access Protocol; communication protocol used in object-oriented architectures
SSCC	Serial Shipping Container Code (EAN 128 standard); equivalent to German NVE
SST	Smart & Secure Tradelanes; US initiative for using RFID technology to increase security in global logistics processes
STANAG	Standard NATO Agreement with reference to the EAN 128 standard
Tag	Abbreviated form of 'RFID tag'
TCP/IP	Transmission Control Protocol/Internet Protocol; Internet data transmission standard
TID	Transponder Identification Number; a unique transponder number stored in the transponder by the manufacturer
Transponder	Coinage formed from 'transmitter' and 'responder' (receiver); abbreviated form of 'RFID transponder' and synonym for 'RFID tag'
UCC	US standardization organization; merged into GS1
UII	Unique Item Identifier; a 96-bit EPC conforming to the EPC Gen 2 specification and the ISO/IEC 18000-6 and 18000-6c standards

UML	Unified Modelling Language; a development language for software systems
UMTS	Universal Mobile Telecommunications System; a third-generation (3G) mobile telephone technology
UPC	Barcode standard used in North America (12 digits); counterpart to EAN
URL	Uniform resource locator; a WWW address on the Internet
Vicinity	Near-field technology standard for smart cards; ISO 15693; *see also* Proximity
VIN	Vehicle Identification Number; used for motor vehicles, etc.
VPN	Virtual private network; a virtual connection in the Internet for secure transmission of commercially sensitive messages
W3C	World Wide Web Consortium; association of companies for Internet standardization, including XML
WiFi	Wireless Fidelity; designation for a WLAN compliant with the ISO 802.11 standard
WiMax	Worldwide Interoperability for Microwave Access; a standard for supraregional WiFi networks
Wireless label	Colloquial term for 'RFID tag'; 'wireless' refers to use of radio signals
WLAN	Wireless LAN; *see* WiFi *and* WiMax
WMS	Warehouse management system; a component of ERP
WSDL	Web Services Description Language
XI	Exchange Infrastructure; SAP software interface
XML	Extended Markup Language; Internet standard for formal description of data and document

Index

Java applets, 140
Java servlets, 140
JIS (just in sequence), 56, 109
JIT (just in time), 56, 108

K
Kanban containers, 79
Key economic figures, 91
Key figures, 91
Key technologies, 16
Kill command, 166, 174, 213

L
Label, 213, 216
Laboratory studies, 183
Labour rates, 89
Laendmarks project, 113
LBA (Luftfahrtbundesamt), 114
Legacy systems, 21, 127
Legitimate market players, 26
Limitation of use to specific
    purposes, 201
Load carriers, 192
Local positioning system, 229
Locational privacy, 210, 217, 218
Logistics Department (FLog) of the
    University of Dortmund, 178
Logistics service provider, 67
    core tasks, 67
Logistics
    flexible and transparent, 249
    of mobile telephones, 245
    strategies, 51
LPS (Local Positioning System), 229
Lufthansa, 115, 240
Luxury consumer goods, 26

M
Magnetic coupling, 166, 173
Main process, 84
Mainframe, 33
Maintenance activity, 225
Maintenance costs, 244
Maintenance management, 80
Maintenance process, 225
Managed services, 158
Mandated requirements, 158
Market barriers, 44
Market maturity, 42
Marks & Spencer, 29

MAS (multiagent system), 137, 145
Mass customization, 57
Mass individualization, 57
Master control system, 144
Material flow control, 146
Material flow system, 148
Mechanical stress, 189
Media break, 9
Media hype, 9
Merchandise flow process, 249
MES (manufacturing execution
    system), 109
Mesh boxes, 77
Meshed networks, 194
Metal envelopes, 214
Metro Group, 2, 29, 31, 251
Microsensys, 224
Middleware, 190, 251
Migration steps, 76
Miniaturization, 33
MIT (Massachusetts Institute of
    Technology), 29
Mobile devices, 33
Mobile RFID reader, 227
Mobile telephone logistics, 245
Mobile telephone tracking system,
    247
MRO (maintenance, repair and
    overhaul), 114
Multi-optional behaviour, 54

N
Net present value, 91
NFC (near-field communication),
    176, 178, 231, 234
NFC Forum, 176
NFC mobile phone, 153, 170
Notice, 215

O
Object analysis, 182
Object website, 97
Object-oriented technology, 125
Obligation to provide information,
    general, 203
Obstacles to the use of RFID, 42
Odette, 111
OECD, 26
    guidelines, 211
On-demand services, 155, 158

ONS (Object Naming Service),
    100, 160
ONS server, 100
Open-loop, 44
Operating costs, 44, 89
Operational service provider, 67
Order batching, 54
Order-to-payment process, 50
Orientation dependence, 185
Out-of-stock situation, 61
Outsourcing, 156

P
Pacemaker technologies, 15
Packaged goods, 71
Packaging industry, 68
Packaging materials, 71
Packaging process costs, 72
Packing aids, 71, 77
Pallet level, 70
Paradigm shift, 40
Parameter variation analysis, 92
Passenger seat Quick-Change, 240
Passive tags, 168
Payment terminal, contactless, 234
Peak of expectations, 14
Pedigree, 27
Performance limits, 43
Personal data, 211, 212, 217, 218
Personal use profile, 204
Person-related data, 201
Pervasive computing, 33
Pharmaceutical industry, 116
Pioneer, 214, 217, 219
PLC (programmable logic
    controller), 170
PLM (product lifecycle
    management), 97
PML (Programmable Markup
    Language), 154
Polymer chips, 38
Polymer tags, 150
Polytronics, 38
Portable reader, 170
POS processes, 200
Positioning system, local, 229
Potential savings, 82
PPC (product planning and
    control), 50
Prescription drugs, 214, 215